AN INTRODUCTION TO
EXCEL VBA PROGRAMMING
with Applications in Finance and Insurance

AN INTRODUCTION TO
EXCEL VBA PROGRAMMING
with Applications in Finance and Insurance

GUOJUN GAN

University of Connecticut
Storrs

CRC Press is an imprint of the
Taylor & Francis Group, an **informa** business

A CHAPMAN & HALL BOOK

CRC Press
Taylor & Francis Group
6000 Broken Sound Parkway NW, Suite 300
Boca Raton, FL 33487-2742

First issued in paperback 2019

ISBN-13: 978-1-138-19715-2 (hbk)
ISBN-13: 978-0-367-26128-3 (pbk)

Visit the Taylor & Francis Web site at
http://www.taylorandfrancis.com

and the CRC Press Web site at
http://www.crcpress.com

To all my students.

Contents

List of Figures

List of Tables

Preface

Visual Basic for Applications (VBA) is a programming language created by Microsoft® that can be used to automate operations in Microsoft Excel, which is perhaps the most frequently used computer software program for manipulating data and building models in banks and insurance companies. One advantage of VBA is that it enables you to do anything that you can do manually in Excel and do many things that Excel does not allow you to do manually. As a powerful tool, VBA has been used by actuaries and financial analysts to build actuarial and financial models.

In the spring of 2016, I was assigned to teach the course "Math3550: Programming for Actuaries," which is taken primarily by junior and senior undergraduate students majoring in actuarial science at the University of Connecticut. This course explores how an actuary uses computers to solve common actuarial problems and teaches students how to design, develop, test, and implement programs using Microsoft Excel with VBA. Since existing books on Excel VBA do not have exercises or applications related to actuarial science, I started to write lecture notes for this course. This textbook has grown out of those lecture notes.

This textbook has been written for undergraduate students majoring in actuarial science who wish to learn the basic fundamentals and applications of Excel VBA. In doing so, this book does not assume that readers have any prior programming experience. This book will also be of use to actuaries and financial analysts working in insurance companies and banks who wish to learn Excel VBA.

This textbook is divided into two parts: preliminaries of Excel VBA programming and some applications of VBA in finance and insurance. The preliminaries covered in the first part include how to run VBA programs, modules, best practices of VBA coding, the Excel object model, variables, control statements, functions, and error handling, among many other things. The applications of VBA introduced in the second part include generating regular payment schedules, bootstrapping yield curves, creating risk-neutral scenarios, pricing a guarantee embedded in a variable annuity contract, how to connect to databases, and object-oriented programming in VBA.

The best way to learn programming is by doing. I encourage readers to practice the VBA code presented in the book. The book also contains many exercises. Sample solutions of some exercises are given in the appendix of this book. Readers should explore the exercises before looking at the solutions.

Finally, I take this opportunity to express my thanks to my students, friends, and colleagues from the University of Connecticut, who have read and provided valuable feedback on the draft of this book.

Guojun Gan
Storrs, Connecticut, USA

Part I

VBA Preliminaries

1

Introduction to VBA

VBA refers to Visual Basic for Applications and is a programming language created by Microsoft® to automate operations in Microsoft Office® applications such as Excel®, Access®, Word®, PowerPoint®, and Outlook®. VBA is a powerful tool that enables you to do whatever you need to do in your job. With VBA, you can do anything that you can do manually and do many things that Excel® does not allow you to do manually. In particular, VBA allows you to

- automate a recurring task.
- automate a repetitive task.
- run a macro automatically if an event occurs.
- build custom functions (i.e., user-defined functions).
- build a customized interface.
- manipulate files and folders.
- manipulate Microsoft Office® applications.
- work with Windows® by calling its Application Programming Interfaces (APIs).
- work with other applications through Dynamic-Link Libraries (DLLs).

In this chapter, we introduce some basic concepts of VBA programming in Excel®. After studying this chapter, readers will be able to

- run VBA macros or programs.
- understand regular and class modules.
- use the Excel® macro record to create VBA macros.

1.1   Getting Started

VBA is a programming language built into Microsoft® applications. If the computer you use has the Office® suite or an individual application, then you can use VBA in that computer. For example, if the computer has Microsoft Excel®, then you can use VBA in the computer.

Since VBA is a programming language built into Microsoft applications, you cannot use VBA to create standalone applications. VBA programs, also referred to as macros, exist within a host application, such as Excel and the workbook containing the VBA code. To run a VBA macro in an Excel workbook, we must open the workbook.

To create a VBA macro in Excel, we first need to open the Visual Basic Editor. In a Windows® computer, for example, we can open the Visual Basic Editor by clicking the icon "Visual Basic" in the left side of the tab "DEVELOPER."[1] Figure 1.1 shows a screen shot of the Excel 2013 interface in a Windows® machine. The Visual Basic Editor in a Windows® machine is shown in Figure 1.2.

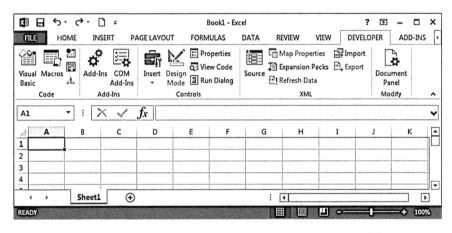

Figure 1.1: Excel 2013 interface in a Windows computer

If you use a Mac® computer, you can open the Visual Basic Editor by clicking the menu item "Tools/Macro/Visual Basic Editor" (see Figure 1.3). You can also click the "Editor" icon in the "Developer" tab to open the Visual Basic Editor. If you do not see the "Developer" tab in your version of Excel,

---

[1]New Ribbon interface has been used since Microsoft Office version 2007. In old versions of Excel, you can open the Visual Basic Editor from the traditional menus.

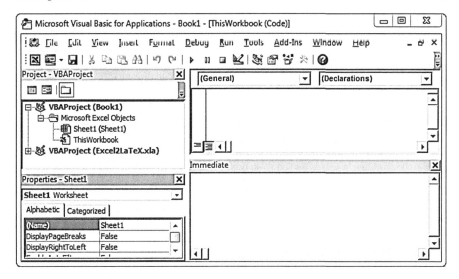

Figure 1.2: Visual Basic Editor in a Windows computer

you can show the tab by clicking "Excel/Preferences.../Ribon" and selecting "Developer" in the Ribon dialog, Mac® shown in Figure 1.4. The Visual Basic Editor in a computer looks different than that in a Windows® machine. Figure 1.5 shows the Visual Basic Editor in a Mac® computer.

There is a shortcut to open the Visual Basic Editor if you do not want to click menu items to open it. In a Windows® machine, you can open Excel and hit "Alt+F11" (i.e., hold the "Alt" key and press the "F11" key) to open the Visual Basic Editor. On a Mac® computer, you can open Excel and hit "fn+option+F11" (i.e., hold the "fn" and "option" keys and press the "F11" key) to open the Visual Basic Editor.

To demonstrate how VBA works in Excel, we can type the following VBA code in the Visual Basic Editor:

```
1  Sub HelloWorld()
2      ThisWorkbook.Worksheets(1).Cells(1, 1) = "Hello
           World!"
3  End Sub
```

The code is contained in a *sub* procedure, which starts with the keyword *sub* and ends with the keyword *End Sub*. The name of the *sub* procedure is "HelloWorld." Line 2 is the VBA command that writes "Hello World!" to the first cell of the first worksheet of the current workbook. Note that VBA

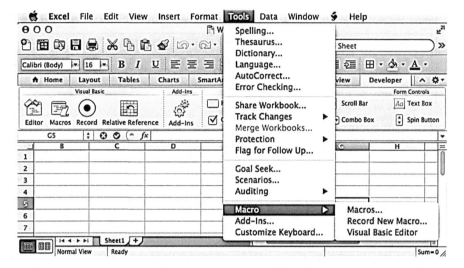

Figure 1.3: Excel 2010 interface on a Mac computer

code is case insensitive. The Visual Basic Editor changes the first letter of a keyword to uppercase.

There are several ways to run the VBA macro. The first way to run the VBA macro is to click the VBA code and press the key "F5." After the code is executed, the content of the first cell of the first worksheet will be changed to "Hello World!"

A second way to run the VBA macro is to open the Macro dialog by clicking "Macros" button in the left side of the "DEVELOPER" tab. On a Mac computer, you can open the macro dialog by clicking "Tools/Macro/- Macros." The macro dialog is shown in Figure 1.6. You can select the macro and click the "Run" button to run the macro. The Macro dialog also allows you to set a shortcut key to your macro (See Exercise 1.1).

────────────────────────────◆────────────────────────────

Exercise 1.1. Suppose that a workbook contains at least two worksheets.

(a) Write a VBA sub procedure named HelloWorld2 that puts "Hello" in the first cell in the first row of the second worksheet and "World" in the second cell in the first row of the second worksheet.

(b) Assign a shortcut key ("Ctrl+Shift+A" on a Windows machine) to the sub procedure and hit the shortcut key.

────────────────────────────◆────────────────────────────

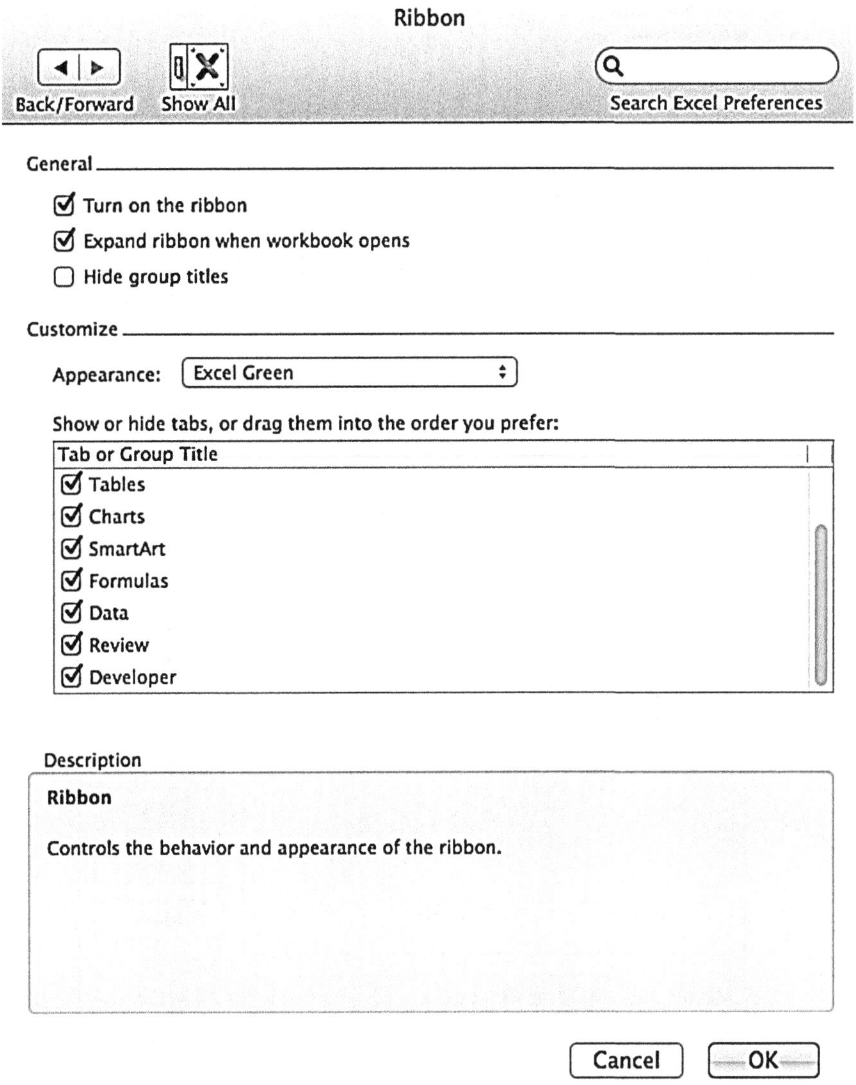

Ribbon

◀	▶	🖺🗙		Q _____

Back/Forward   Show All   Search Excel Preferences

General _____

    ☑ Turn on the ribbon

    ☑ Expand ribbon when workbook opens

    ◯ Hide group titles

Customize _____

    Appearance: [Excel Green   ⬍]

    Show or hide tabs, or drag them into the order you prefer:

Tab or Group Title	
☑ Tables	
☑ Charts	
☑ SmartArt	
☑ Formulas	
☑ Data	
☑ Review	
☑ Developer	

Description

Ribbon

Controls the behavior and appearance of the ribbon.

[Cancel]   [OK]

Figure 1.4: Ribbon dialog in Excel 2010 on a Mac computer

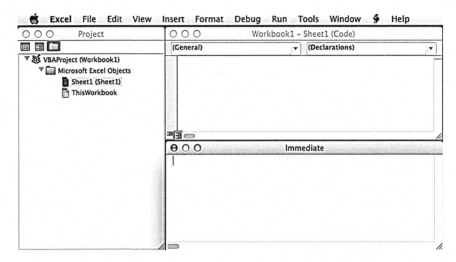

Figure 1.5: Visual Basic Editor on a Mac computer

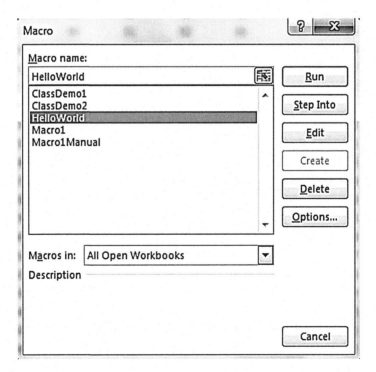

Figure 1.6: Macro dialog

1.2 Modules

A module is a container for VBA code. A module can contain declarations and procedures. An Excel application can call VBA code contained in one or more modules to perform specified tasks.

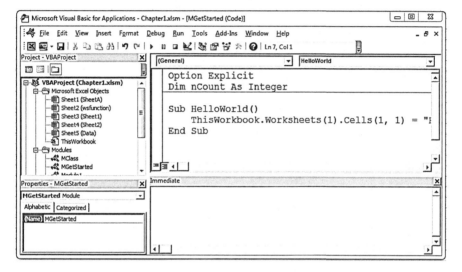

Figure 1.7: A module displayed in the Visual Basic Editor

Figure 1.7 shows a module displayed in the Visual Basic Editor. The name of the module is `MGetStarted`. You can change the name of the module in the Properties window, which is shown in the bottom left of the figure.

As we can see from the figure, the module contains two declarations and one procedure. The two declarations are:

```
1 Option Explicit
2 Dim nCount As Integer
```

The first statement contains two keywords: `Option` and `Explicit`. This statement will force the explicit variable declaration, that is, you have to declare a variable explicitly before using the variable. We will see some examples of this statement in Section 3.1. The second statement declares an integer variable named `nCount`.

A module can contain two types of procedures: sub procedures and function procedures. A sub procedure contains VBA code to perform some operations and does not return a particular result. A function procedure performs some operations and returns a particular result. We will introduce function procedures in Section 5.1.

To create a module, we can right click the Modules folder in the Project window and then click "Insert/Module" (See Figure 1.8). The Project window is shown in the top left of Figure 1.8. On a Mac computer, you need to hold the Control key and click the Modules folder.

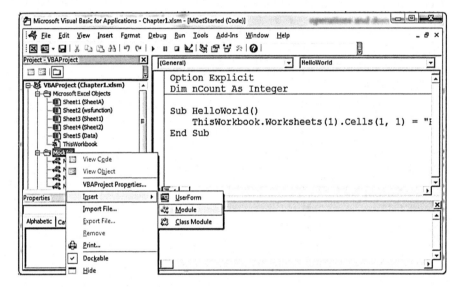

Figure 1.8: Inserting a module in the Visual Basic Editor

There are two types of modules: standard modules and class modules. If a module contains procedures that are not associated with any particular object, then module is a standard module. If a module is associated with a particular object, then it is a class module.

Class modules are related to object-oriented programming and are different from standard modules in the way their data is stored. Only one copy of a standard module's data is stored. For example, if one part of your program changes a public variable in a standard module and another part of your program subsequently reads that variable, then it will get the same value. Class module data is stored separately for each object. We will introduce how to use class modules in Section 1.3.

Modules have the following advantages:

- Modules can be exported and imported into other workbooks.

- Modules can be organized in a logical way. For example, you can dedicate a separate module for data access, spreadsheet formatting, and utilities.

Exercise 1.2. Create a standard module and change its name to MTest.

1.3 VBA Classes

VBA supports object-oriented programming. A class is a core concept of object-oriented programming. In this section, we introduce how to define and use classes in VBA.

In VBA, a class is defined in a class module. To show how to create a class in VBA, we first insert a class module named CStudent and type the following code in the class module:

```
 1 Option Explicit
 2
 3 Dim sFirstName As String
 4 Dim sLastName As String
 5 Dim sGrade As String
 6
 7
 8 Public Property Get FirstName() As String
 9     FirstName = sFirstName
10 End Property
11
12 Public Property Let FirstName(ByVal FirstName As
      String)
13     sFirstName = FirstName
14 End Property
15
16 Public Property Get LastName() As String
17     LastName = sLastName
18 End Property
19
20 Public Property Let LastName(ByVal LastName As String
      )
```

```
21      sLastName = LastName
22 End Property
23
24 Public Property Get Grade() As String
25      Grade = sGrade
26 End Property
27
28 Public Property Let Grade(ByVal Grade As String)
29      sGrade = Grade
30 End Property
31
32 Public Sub ShowInfo()
33      MsgBox sFirstName & " " & sLastName & "'s grade
           is " & sGrade
34 End Sub
```

The name of the class is the same as the name of the class module. From the above code, we see that the class CStudent has three properties: FirstName, LastName, and Grade. The class module also contains six property procedures to set and get the values stored in the properties. The sub procedure ShowInfo is used to display the property values in a message box.

To show how to use the class CStudent, we consider the following sub procedure:

```
 1 Sub ClassDemo1()
 2      Dim cS1 As CStudent
 3      Dim cS2 As CStudent
 4
 5      Set cS1 = New CStudent
 6      cS1.FirstName = "Graig"
 7      cS1.LastName = "Moyle"
 8      cS1.Grade = "A"
 9
10      Set cS2 = New CStudent
11      cS2.FirstName = "Violet"
12      cS2.LastName = "Keen"
13      cS2.Grade = "B"
14
15      cS1.ShowInfo
16      cS2.ShowInfo
17 End Sub
```

In the above code, we create two objects of the class CStudent by using the keyword New. We assign the two objects to two variables by using the keyword Set. Then we call the class's method ShowInfo to display the property values of the two objects.

———————————————————●———————————————————

Exercise 1.3. Create a VBA class named CBasket with the following two properties: iCount and dPrice. The first property stores how many items in the basket. The second property stores the price of each item. The class also has a method named CalculateValue to calculate the total value of the basket. Then write a sub procedure to test the class.

———————————————————●———————————————————

1.4 The Excel Macro Recorder

In previous sections, we created Excel macros manually. Using the Excel macro recorder is another way to create macros. In this section, we introduce how to use the Excel macro recorder to create macros, and the advantages and drawbacks of the Excel macro recorder.

To record a macro using the Excel macro recorder, we first click the "Record Macro" button in the left side of the "Developer" tab (See Figure 1.1). A dialog named "Record Macro" will pop up. Figure 1.9 shows the dialog on a Windows machine. We can change the name, create a shortcut key, select where to store the macro, and input the description in the dialog. In our example, we use the default settings and just write some description in the Description textbox. Then we click the "OK" button to start recording. We select the cell "D3," type "Hello," change the cell borders to bold lines, and select the cell "D5." Then we click the stop button located in the bottom left of the Excel window.

The recorded macro is saved to a new module of this workbook. In our example, the recorded macro was saved to Module2. If we click Module2 in the Visual Basic Editor, we see the following sub procedure:

```
1  Sub Macro1()
2  '
3  '  Macro1 Macro
4  '  Excel macro recorder demo
5  '
6
```

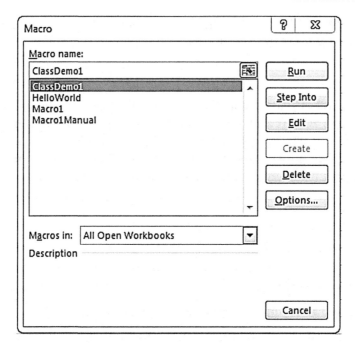

Figure 1.9: The Record Macro dialog

```
 7
 8    Range("D3").Select
 9    ActiveCell.FormulaR1C1 = "Hello"
10    Range("D3").Select
11    Selection.Font.Bold = True
12    Selection.Borders(xlDiagonalDown).LineStyle =
          xlNone
13    Selection.Borders(xlDiagonalUp).LineStyle =
          xlNone
14    With Selection.Borders(xlEdgeLeft)
15        .LineStyle = xlContinuous
16        .ColorIndex = 0
17        .TintAndShade = 0
18        .Weight = xlMedium
19    End With
20    With Selection.Borders(xlEdgeTop)
21        .LineStyle = xlContinuous
22        .ColorIndex = 0
23        .TintAndShade = 0
24        .Weight = xlMedium
25    End With
```

```
26    With Selection.Borders(xlEdgeBottom)
27        .LineStyle = xlContinuous
28        .ColorIndex = 0
29        .TintAndShade = 0
30        .Weight = xlMedium
31    End With
32    With Selection.Borders(xlEdgeRight)
33        .LineStyle = xlContinuous
34        .ColorIndex = 0
35        .TintAndShade = 0
36        .Weight = xlMedium
37    End With
38    Selection.Borders(xlInsideVertical).LineStyle =
          xlNone
39    Selection.Borders(xlInsideHorizontal).LineStyle =
          xlNone
40    Range("D5").Select
41 End Sub
```

In the above code, Lines 2–5 and 7 start with a single quote. These lines are comments in VBA code.

The Excel macro recorder is useful in that it helps us figure out what VBA commands we can use to do certain things, such as formatting a range. However, the Excel macro recorder has some drawbacks. First, the Excel macro recorder cannot create functions for us. It only produces sub procedures. Second, the VBA code created by the recorder is not compact. It tends to record more statements than we need. For example, we can write the following sub procedure that does the same thing as the recorded macro:

```
1 Sub Macro1Manual()
2    ' This macro does the same thing as Macro1
         recorded by the Excel macro recorder
3    Range("D3").Select
4    ActiveCell.Value = "Hello"
5    ActiveCell.Font.Bold = True
6    Selection.Borders.LineStyle = xlContinous
7    Selection.Borders.Weight = xlMedium
8 End Sub
```

The macro written manually is much shorter than the recorded macro.

Exercise 1.4. Suppose that the range "A1:B2" contains four numbers. Use the Excel macro recorder to create a macro that does the following operations: select the range "A1:B2," copy and paste the range to the range "F1:G2," and change the number format of the range "F1:G2" to a percentage with two decimal places.

Exercise 1.5. Use the Excel macro recorder to record the following steps: open the file "USLifeTable1999-2001Male.xls" by clicking "File/Open," select the range "A25:B134," copy and paste the range to the range "A1:B110" of a worksheet in your own workbook, and close the workbook "USLifeTable1999-2001Male.xls."

1.5 Summary

In this chapter, we introduced some basic concepts of VBA programming. In particular, we introduced how to run VBA programs. We also introduced VBA modules, including standard modules and class modules. Finally, we introduced the Excel macro recorder, which is a useful tool to create VBA commands.

There are quite a few books on VBA programming. For a brief history of VBA, readers are referred to Urtis (2015). Readers who are beginners of VBA programming are referred to Martin (2012), Walkenbach (2013b), and Urtis (2015). Advanced readers are referred to Getz and Gilbert (2001) and Bovey et al. (2009). For financial computing with Excel and VBA, readers are referred to Lai et al. (2010).

2

Excel Objects

Programming in Visual Basic for Applications (VBA) is related to manipulating the objects that make up Excel® and other Office applications. In this chapter, we introduce the Excel object model and some commonly used Excel objects. After studying this chapter, readers will be able to

- understand the Excel object model.
- use the properties and methods of some common objects, such as `Application`, `Workbook`, `Worksheet`, `Range`, and `WorksheetFunction`.

2.1   Excel Object Model

Object-oriented programming (OOP) is a style of programming based on the use of objects and their interactions. An object is a data structure that contains properties and methods. Properties of an object are used to store data. Methods of an object are used to access and modify the data stored in properties. An object can contain and interact with other objects.

Figure 2.1 shows a hierarchy of Excel objects. The `Application` object (i.e., Excel itself) is at the top of the hierarchy and contains other objects. The `Workbook` object and the `WorksheetFunction` object are examples of objects contained in the `Application` object. The `Workbook` object also contains other objects, such as `Worksheet`, `Name`, and `VBProject`. The `Worksheet` object contains `Name`, `Range`, `Cells`, and other objects. The `Range` object can also contain the `Cells` object.

A special Excel object is the collection object, which contains objects of the same type. For example, the `Workbooks` object contains all currently open `Workbook` objects. The `Worksheets` object contains all `Worksheet` objects contained in a particular `Workbook` object.

Figure 2.1 shows only a subset of the Excel objects. To see the complete list of Excel objects and their members, we can use the Object Browser

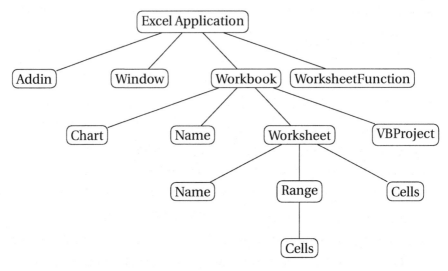

Figure 2.1: The Excel object hierarchy

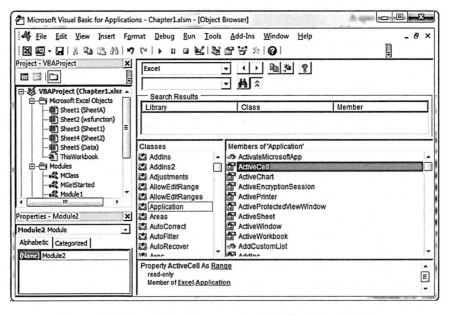

Figure 2.2: The Object Browser in the Visual Basic Editor

shown in Figure 2.2. We can open the Object Browser by clicking the menu item "View/Object Browser."

We use the dot operator to access an object in the object hierarchy. To reference a single object from a collection, we use the object's name or index number. In the following sub procedure, for example, we first assign the text "A String" to the second cell in the first row of the first sheet. Then we use the VBA function Debug.Print to display the value of that cell and the name of the current workbook in the Immediate window.

```
1 Sub AccessObject()
2     Application.ActiveWorkbook.Sheets(1).Cells(1, 2).
          Value = "A String"
3     Debug.Print Application.ActiveWorkbook.Sheets("
          Sheet1").Cells(1, 2).Value
4     Debug.Print Application.ActiveWorkbook.Name
5 End Sub
```

If the sub procedure is executed, we see the following output in the Immediate window:

```
1 A String
2 Book1
```

If the Immediate window is not shown in the Visual Basic Editor, you can open the Immediate window by clicking the menu item "View/Immediate Window."

We can simplify the reference to an object in the object hierarchy by omitting references to the assumed objects. If an object is the currently active object, we can omit the reference to the object. For example, we can rewrite the sub procedure AccessObject as follows:

```
1 Sub AccessObject2()
2     Sheets(1).Cells(1, 2).Value = "A String"
3     Debug.Print ActiveWorkbook.Sheets("Sheet1").Cells
          (1, 2).Value
4     Debug.Print ActiveWorkbook.Name
5 End Sub
```

In this version, we omit the reference to the object Application.

Exercise 2.1. Write a sub procedure named DisplayCount to display the number of open workbooks and the number of worksheets of the current workbook in the Immediate window.

Exercise 2.2. Write a sub procedure named `CountRowColumn` to display the number of rows and the number of columns of a worksheet in your Excel application. You can use the objects `Rows`, `Columns`, and `Count` in your code.

As we mentioned before, an object contains properties and methods. A method can modify an object's properties or make the object do something. The following sub procedure shows how to use the VBA function `Copy` to copy the content of a cell to another cell.

```
1 Sub CopyValue()
2     ThisWorkbook.WorkSheets(1).Activate
3     Cells(2, 2).ClearContents
4     Debug.Print Cells(1, 1).Value
5     Debug.Print Cells(2, 2).Value
6     Cells(1, 1).Copy Cells(2, 2)
7     Debug.Print Cells(2, 2).Value
8 End Sub
```

In the above code, we first activate the first worksheet by calling the method `Activate`. Then we call the method `ClearContents` to clear the content of the cell at the second row and the second column. In Lines 4–5, we call the VBA function `Debug.Print` to display the contents of the two cells in the Immediate window. In Lines 6–7, we copy the content of the first cell to the second cell and display the content of the second cell again. If we execute the above code, we see the following output in the Immediate window:

```
1 Hello World!
2
3 Hello World!
```

Exercise 2.3. Suppose that a workbook contains at least two worksheets. Write a sub procedure named `CopyValue2` to copy the content of the cell "A1" in the first worksheet to the cell "B2" in the second worksheet.

2.2 Application Object

As mentioned in the previous section, the Application object is the top-most object in the Excel object model. The Application object contains application-wide settings and options and top-level objects and methods.

Table 2.1: Commonly Used Properties of the Application Object

Property	Description
ActiveCell	Returns a Range object that represents the active cell
ActiveSheet	Returns an object that represents the active sheet
ActiveWorkbook	Returns a Workbook object that represents the workbook in the active window
ThisWorkbook	Returns a Workbook object that represents the workbook where the current macro code is running
DisplayAlerts	True if Excel displays certain alerts and messages while a macro is running. Read/write boolean Read/write boolean
Columns	Returns a Range object that represents all the columns on the active worksheet
Rows	Returns a Range object that represents all the rows on the active worksheet
Name	Returns a string that represents the name of the object
Sheets	Returns a Sheets collection that represents all the sheets in the active workbook
Workbooks	Returns a Workbooks collection that represents all the open workbooks
FileDialog	Returns a FileDialog object that represents an instance of the file dialog

The Application object contains many properties and methods. A list of commonly used properties is shown in Table 2.1. The following sub procedure illustrates how we can refer to some of these properties:

```
1  Sub ApplicationDemo()
2      Debug.Print Application.ActiveWorkbook.Name
3      Debug.Print Application.ActiveSheet.Name
4      Debug.Print Application.ActiveCell.Address
5      Debug.Print Application.Name
6  End Sub
```

Executing the above code produces the following output in the Immediate window:

```
1  Chapter1.xlsm
2  Sheet1
3  $D$1
4  Microsoft Excel
```

Table 2.2: Commonly Used Methods of the `Application` Object

Method	Description
Evaluate	Converts a Microsoft Excel name to an object or a value
FindFile	Displays the Open dialog box
GetOpenFilename	Displays the standard Open dialog box and gets a file name from the user without actually opening any files
InputBox	Displays a dialog box for user input and returns the information entered in the dialog box
OnKey	Runs a specified procedure when a particular key or key combination is pressed
OnTime	Schedules a procedure to be run at a specified time in the future
Run	Runs a macro or calls a function

The following sub procedure shows how to use the methods `InputBox` and `Evaluate`:

```
1  Sub ApplicationDemo2()
2      Debug.Print Application.InputBox(prompt:="Input a
           cell address", Type:=2)
3      Application.ActiveSheet.Range("B10").Value = 3.14
4      Debug.Print Application.Evaluate("B10").Value
5  End Sub
```

If we execute the above code, an input box will show up on the screen. If we type "A1" (without the double quotes) in the input box, we will see the following output in the Immediate window:

```
1  A1
2   3.14
```

The function `Evaluate` returned a `Range` object from the reference "B10."

Exercise 2.4. Read the Application.InputBox Method (Excel) at `https://msdn.microsoft.com/en-us/library/office/ff839468.aspx` and write a sub procedure named `InputBoxDemo` to do the following: use the method `InputBox` to let users input a number and then display the sum of the input and 5 in the Immediate window.

Exercise 2.5. Read the Application.OnKey Method (Excel) at `https://msdn.microsoft.com/en-us/library/office/ff197461.aspx` and write a sub procedure named `OnKeyDemo` that assigns the key combination "Shift + Ctrl + A" to the following sub procedure:

```
1 Sub HelloWorld()
2     MsgBox "Hello World!"
3 End Sub
```

For a complete list of members of the `Application` object and how to use them, readers can read the MSDN document at `https://msdn.microsoft.com/en-us/library/office/ff194565.aspx`. For more information about the `Application` object, readers are referred to the Developer's Guide to the Excel 2010 Application Object at `https://msdn.microsoft.com/en-us/library/office/gg192737%28v=office.14%29.aspx`.

2.3 Workbook Objects

A `Workbook` object represents a Microsoft Excel workbook. A `Workbook` object is a member of the `Workbooks` collection, which contains all the `Workbook` objects currently open in Microsoft Excel.

In addition to the `Workbooks` property, the `Application` object has other properties related to workbooks: `ThisWorkbook` and `ActiveWorkbook`. We can get a `Workbook` object from the `Workbooks` collection. We can also get a `Workbook` object from the properties `ThisWorkbook` and `ActiveWorkbook`. The following sub procedure shows how to refer to a workbook using three approaches:

```
1  Sub  WorkbookDemo1()
2        Debug.Print  Application.ThisWorkbook.FullName
3        Debug.Print  Application.ActiveWorkbook.FullName
4        Debug.Print  Application.Workbooks(1).FullName
5  End  Sub
```

Executing the above code produces the following output in the Immediate window:

```
1  Scratch:gan:ResearchU:trunk:book:vbapfi:code:Chapter2
       .xlsm
2  Scratch:gan:ResearchU:trunk:book:vbapfi:code:Chapter2
       .xlsm
3  Scratch:gan:ResearchU:trunk:book:vbapfi:code:Chapter2
       .xlsm
```

The above output was produced by running the code on a Mac® computer. That is why we see colons in the file paths. If we run the code on a Windows® machine, we see the following output:

```
1  C:\Users\Guojun  Gan\Documents\ResearchU\trunk\book\
       vbapfi\code\Chapter2.xlsm
2  C:\Users\Guojun  Gan\Documents\ResearchU\trunk\book\
       vbapfi\code\Chapter2.xlsm
3  C:\Users\Guojun  Gan\Documents\ResearchU\trunk\book\
       vbapfi\code\Chapter2.xlsm
```

From the output, we see that all three approaches refer to the same workbook.

There are several types of Excel workbooks, each of which has a different file extension. Table 2.3 presents a list of Excel file extensions. Except for the extension .xll, all file extensions were introduced with Excel 2007. Using these extensions, we can save a workbook without macros or with macros. We can also save a workbook as a template without macros or with macros. Add-in workbooks are workbooks that contain code. Using add-in workbooks allows the user to separate code from the users' data.

Exercise 2.6. Read the Application.ThisWorkbook Property (Excel) at https://msdn.microsoft.com/en-us/library/office/ff193227.aspx and examine whether the following statements are true or false:

Table 2.3: File Extensions of Excel Files

Extension	Description
.xlsx	Workbook
.xlsm	Macro-enabled workbook; same as .xlsx but may contain macros and scripts
.xltx	Template
.xltm	Macro-enabled template; same as .xltx but may contain macros and scripts
.xlam	Add-in workbook

(a) `ThisWorkbook` is the only way to refer to an add-in workbook from inside the add-in itself.

(b) `ThisWorkbook` always returns the workbook in which the code is running.

(c) `ThisWorkbook` always is the same as `ActiveWorkbook`.

Table 2.4: Commonly Used Properties of a `Workbook` Object

Property	Description
`ActiveChart`	Returns a `Chart` object that represents the active chart
`ActiveSheet`	Returns an object that represents the active sheet
`Charts`	Returns a `Sheets` collection representing all the chart sheets
`FullName`	Returns the name of the object, including its path
`IsAddin`	True if the workbook is running as an add-in
`Name`	Returns the name of the object
`Path`	Returns a string representing the complete path to the workbook
`Sheets`	Returns a `Sheets` collection that represents all the sheets in the workbook
`Worksheets`	Returns a `Sheets` collection that represents all the worksheets in the specified workbook

A `Workbook` object has many properties. Table 2.4 shows a list of some commonly used properties of a `Workbook` object. We already saw how to

refer to the FullName property of a Workbook object. The following sub
procedure shows how to refer to other properties:

```
1 Sub  WorkbookDemo2()
2       Debug.Print  ThisWorkbook.Charts.Count
3       Debug.Print  ThisWorkbook.Sheets.Count
4       Debug.Print  ThisWorkbook.Worksheets.Count
5       Debug.Print  ThisWorkbook.IsAddin
6 End  Sub
```

Executing the above code produces the following output:

```
1  0
2  1
3  1
4 False
```

The output shows that the current workbook contains one worksheet and
no charts. The workbook is not an add-in workbook.

Table 2.5: Commonly Used Methods of a Workbook Object

Method	Description
Activate	Activates the first window associated with the workbook
Close	Closes the object
Save	Saves changes to the specified workbook
SaveAs	Saves changes to the workbook in a different file

Table 2.5 shows a list of some commonly used methods of a Workbook
object. Using these methods we can activate, close, and save a workbook.
For example, executing the following sub procedure will save changes and
then close the current workbook:

```
1 Sub  WorkbookDemo3()
2       ThisWorkbook.Save
3       ThisWorkbook.Close
4 End  Sub
```

Note that a Workbook object does not have the Open method.

Exercise 2.7. Write a sub procedure named WorkbookDemo4 to save the workbook "Chapter2.xlsm" to a new file called "Chapter2Copy.xlsm" in the same folder.

Table 2.6: Two Properties of the Workbooks Object

Property	Description
Count	Returns the number of objects in the collection
Item	Returns a single object from the collection

Table 2.7: Some Methods of the Workbooks Object

Method	Description
Add	Creates a new workbook, which becomes the active workbook
Close	Closes all workbooks
Open	Opens a workbook

The Workbooks object does not have many properties or methods. Table 2.6 and Table 2.7 present some properties and methods of the Workbooks object. From Table 2.7, we see that the Workbooks object provides methods to create and open a workbook. For example, we can open the workbook "Chapter2Copy.xlsm" (See Exercise 2.7) as follows:

```
1 Sub WorkbookDemo5()
2     Debug.Print ActiveWorkbook.Name
3     Workbooks.Open "Chapter2Copy.xlsm"
4     Debug.Print Workbooks.Count
5     Debug.Print ActiveWorkbook.Name
6 End Sub
```

Executing the above code produces the following output:

```
1 Chapter2.xlsm
2  2
3 Chapter2Copy.xlsm
```

From the output we see that the active workbook was changed to the workbook opened by the Open method.

2.4 Worksheet Objects

A Worksheet object represents a worksheet and is a member of the Worksheets collection and the Sheets collection. The Worksheets collection contains all the Worksheet objects in a workbook. The Sheets collection contains all the sheets in a workbook, including charts and worksheets. The Charts collection contains all the charts in a workbook.

Table 2.8: Commonly Used Properties of a Worksheet Object

Property	Description
Cells	Returns a Range object that represents all the cells on the worksheet
Columns	Returns a Range object that represents all the columns on the active worksheet
Index	Returns the index number of the object within the collection of similar objects
Name	Returns or sets a String value that represents the object name
Next	Returns a Worksheet object that represents the next sheet
Range	Returns a Range object that represents a cell or a Range of cells
Rows	Returns a Range object that represents all the rows on the specified worksheet
Type	Returns an XlSheetType value that represents the worksheet type
UsedRange	Returns a Range object that represents the used range on the specified worksheet

A Worksheet object has many properties and provides many methods. Some commonly used properties and methods are described in Table 2.8 and Table 2.9, respectively. All of these properties are read-only except for the property Name. The following sub procedure shows how to refer to a Worksheet object's properties:

```
1  Sub WorksheetDemo1()
2      Debug.Print ActiveSheet.Index
3      Debug.Print ActiveSheet.Name
4      Debug.Print ActiveSheet.UsedRange.Address
5      ActiveSheet.Name = "SheetA"
6      Debug.Print ActiveSheet.Name
```

```
7 End Sub
```

The above code displays the index, name, and used range of the active sheet. Then the code changes the name of the active sheet and displays the name again. Executing the code produces the following output:

```
1   1
2 SheetA
3 $A$1:$G$110
4 SheetA
```

From the output, we see that the used range is "A1:G110." This means that all the cells outside of "A1:G110" are not used. It is possible that some cells in the used range are not used.

Exercise 2.8. What is wrong with the following code?

```
1 Sub WorksheetDemo2()
2     Debug.Print ActiveSheet.Index
3     ActiveSheet.Index = 2
4     Debug.Print ActiveSheet.Index
5 End Sub
```

The following sub procedure shows how to copy and select a worksheet:

```
1 Sub WorksheetDemo3()
2     Debug.Print ActiveSheet.Name
3     ActiveSheet.Copy After:=ActiveSheet
4     Debug.Print ActiveSheet.Name
5     ActiveSheet.Name = "Sheet2"
6     Debug.Print ActiveSheet.Name
7     Sheets("SheetA").Select
8     Debug.Print ActiveSheet.Name
9 End Sub
```

Executing the above code produces the following output in the Immediate window:

```
1 SheetA
2 SheetA (2)
```

Table 2.9: Commonly Used Methods of a `Worksheet` Object

Method	Description
Activate	Makes the current sheet the active sheet
Calculate	Calculates all open workbooks, a specific worksheet, or a specified range on a worksheet
Copy	Copies the sheet to another location in the workbook
Delete	Deletes a worksheet
Move	Moves the sheet to another location in the workbook
Paste	Pastes the contents of the Clipboard onto the sheet
PasteSpecial	Pastes the contents of the Clipboard onto the sheet, using a specified format
SaveAs	Saves changes to the chart or worksheet in a different file
Select	Selects the object

```
3  Sheet2
4  SheetA
```

From the output, we see that the new worksheet becomes the active sheet. The following sub procedure shows how to delete a worksheet:

```
1  Sub WorksheetDemo4()
2      Debug.Print ActiveSheet.Name
3      ActiveSheet.Copy After:=ActiveSheet
4      Debug.Print ActiveSheet.Name
5      ActiveSheet.Delete
6      Debug.Print ActiveSheet.Name
7  End Sub
```

If you run the above code, you will see a prompt dialog. If you click the "Delete" button in the dialog, you will see the following output in the Immediate window:

```
1  SheetA
2  SheetA (2)
3  SheetA
```

From the output, we see that the active sheet changed to the previous worksheet after the active sheet was deleted. If you do not want to see the prompt dialog when deleting a worksheet, you can set the application property DisplayAlerts to be false. For example, we can get rid of the prompt in the following code:

```
1  Sub  WorksheetDemo5()
2      Application.DisplayAlerts = False
3      Debug.Print ActiveSheet.Name
4      ActiveSheet.Copy After:=ActiveSheet
5      Debug.Print ActiveSheet.Name
6      ActiveSheet.Delete
7      Debug.Print ActiveSheet.Name
8      Application.DisplayAlerts = True
9  End  Sub
```

Executing the above code will not show the prompt dialog.

Table 2.10: Some Methods of the Worksheets Collection

Method	Description
Add	Creates a new worksheet, chart, or macro sheet, which becomes the active sheet
Copy	Copies the sheet to another location in the workbook
Delete	Deletes the object
Move	Moves the sheet to another location in the workbook

The Worksheets collection and the Sheets collection have properties and methods that are similar to those of the Workbooks collection. Those properties and methods are shown in Table 2.6 and Table 2.7, respectively. Table 2.10 shows some methods of a Worksheets collection.

Exercise 2.9. Write a sub procedure named WorksheetDemo6 to do the following: create a new worksheet, rename the new worksheet to "Data," and write the index of the new worksheet to the cell "A1" of the new worksheet.

2.5 Range Object

Excel is all about cells. A Range object is a collection of cells and represents a range contained in a Worksheet object. Like other objects, a Range object has properties and methods.

A Range object can contain a single cell or all cells in a worksheet. We can find out the number of cells in a worksheet by multiplying the number of rows and the number of columns (see Exercise 2.2). In Excel 2013, for example, a worksheet contains

$$1048576 \times 16384 = 17,179,869,184$$

cells.

We can refer to a Range object in different ways. Table 2.11 gives a list of expressions that can be used to refer to a Range object. In this table, we only show the expression to refer to a Range object in the ActiveWorksheet object. To refer to a Range object in a worksheet that is different from the active worksheet, we can include the parent object in the reference.

Table 2.11: Various Ways to Refer to a Range Object

Expression	Description
Range("A3:C4")	Cells in Columns A to C and Rows 3 to 4
Range("A1")	The cell in Column A and Row 1
Range("RName")	The range named "RName"
Range("1:1")	Cells in Row 1
Range("A:A")	Cells in Column A
Range("A1:C3,H5:K6")	Cells in noncontiguous ranges
Cells(1,1)	Equivalent to Range("A1")
Range(Cells(3,1), Cells(4,3))	Equivalent to Range("A3:C4")
Range("A1").Offset(2,3)	The cell two rows below A1 and three columns to the right of A1

For example, the active workbook is Chapter2.xlsm, and the workbook USLifeTable1999-2001Male.xls is also open. The second workbook contains the mortality table of the male population in the US (see Figure 2.3). The following sub procedure illustrates how to refer to the range that contains the mortality rates:

```
1  Sub MaleTable()
2      Debug.Print Workbooks("USLifeTable1999-2001Male.
           xls").Worksheets("Sheet1").Range("A25:B134").
           Cells.Count
3  End Sub
```

In the above code, we refer to the workbook and the worksheet that contain the Range object. Executing the sub procedure gives the number of cells contained in the range.

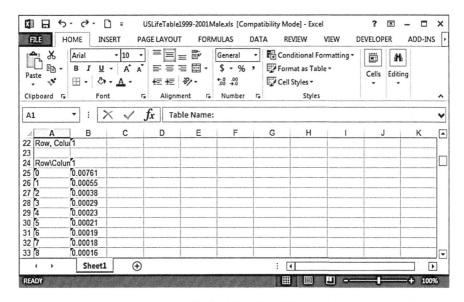

Figure 2.3: A screenshot of the US Life Tables 1999–2001 — Males

From Table 2.11, we see that we can also refer to a Range object by Cells. As pointed out by Walkenbach (2013b), Cells is a property not an object or collection. VBA evaluates a Cells property and returns a Range object. The Cells property takes two arguments. The first argument represents the row index, and the second argument represents the column index. The Cells property provides a useful way to refer to a range if we know the coordinates of the range. Suppose that we want to find the address of the range in the intersection of Rows 10–12 and Columns 20–22. Then we can use the following sub procedure:

```
1  Sub FindAddress()
2      Debug.Print Range(Cells(10, 20), Cells(12, 22)).
           Address
```

```
3 End  Sub
```

Executing the above code gives the following output:

```
1 $T$10:$V$12
```

Exercise 2.10. Write a sub procedure named `FindAddress2` to display the address of the cell in the 20th row and 51st column.

Exercise 2.11. Write a sub procedure named `ReferCell` to display the address of the cell "E10." Use three different ways to refer to the cell "E10."

The `Offset` provides another useful way to refer to a `Range` object relative to another `Range` object. Like the `Cells` object, the `Offset` object also takes two arguments, which represent the number of rows and the number of columns to offset, respectively. For example, we can refer to the cell "B2" as follows:

```
1 Sub  ReferCell2()
2     Debug.Print  Range("A1").Offset(1,  1).Address
3     Debug.Print  Range("E4").Offset(-2,  -3).Address
4 End  Sub|
```

Executing the code produces the following output:

```
1 $B$2
2 $B$2
```

Exercise 2.12. Determine the output of the following VBA code:

```
1 Sub  OffsetRange()
2     Debug.Print  Range("A2:C3").Offset(2,  3).Address
3     Debug.Print  Range("G6:I7").Offset(-2,  -3).Address
4 End  Sub
```

Calculate the output before running the code in the Visual Basic Editor.

Table 2.12: Some Properties of a Range Object

Property	Read-Only	Description
Value	No	The value contained in a cell
Text	Yes	The string displayed in a cell
Count	Yes	The number of cells in a range
Column	Yes	The column index of a cell
Row	Yes	The row index of a cell
Address	Yes	The address of a range in absolute reference
Font	No	Returns a Font object
Interior	No	Returns an Interior object
Formula	No	The formula in a cell
NumberFormat	No	A string that represents the number format

A Range object has many properties. Table 2.12 presents a list of some commonly used properties of a Range object. Some of these properties are read-only. For read-only properties, VBA code can look at their values but cannot change them. Examples of read-only properties include Text and Count. If a property is not read-only, then you can get and set the value of the property.

The following example shows how to set the value of a cell, change the number format, and display the string as shown in the worksheet:

```vba
1 Sub ShowValueText()
2     Range("A1").Value = 198.35
3     Range("A1").NumberFormat = "$#,##0_);($#,##0)"
4     Debug.Print Range("A1").Value
5     Debug.Print Range("A1").Text
6 End Sub
```

Executing the above code gives the following output:

```
1 198.35
2 $198
```

Exercise 2.13. Write a sub procedure named ChangeFont to do the following: Set the value of cell "B2" to the string "World," and change the font to be bold and the font size to be 16.

We can change the background color of a range by using the Interior object. In the following example, we change the background color of cell "C3" to blue and the font color to white:

```
1 Sub ChangeCellColor()
2     Range("C3").Value = "White"
3     Range("C3").Interior.Color = vbBlue
4     Range("C3").Font.Color = vbWhite
5 End Sub
```

If we execute the above sub procedure, cell "C3" will show the word "White" in white and the background color will be changed to blue. Note that vbBlue and vbWhite are VBA constants used to represent colors.

Table 2.13: Some Methods of a Range Object

Method	Description
Select	Select a range of cells
Copy	Copy a range
PasteSpecial	Paste a range from the Clipboard into a specified range
Clear	Delete contents of a range
Delete	Delete a range and fill it up by shifting remaining cells

A Range object also has many methods. Table 2.13 shows some commonly used methods of a Range object. The following piece of code shows how to copy and paste a range:

```
1 Sub CopyRange()
2     Range("A1:C3").Copy
3     Range("D5").PasteSpecial xlPasteAll
4     Debug.Print Selection.Address
5 End Sub
```

After executing the above code, the range "A1:C3" will be copied to the range "D5:F7." Since we specify xlPasteAll, the format of the range is also copied.

Note that a Range object does not have a method called paste. In fact, paste is a method of a Worksheet object. For example, we can copy a range as follows:

```
1 Sub CopyRange2()
2     Range("A1:C3").Copy
3     Range("D5").Select
4     ActiveSheet.Paste
5     Debug.Print Selection.Address
6 End Sub
```

The sub procedure CopyRange and the sub procedure CopyRange2 produce the same result. If we just want to copy values of a range, then we need to use the PasteSpecial method by specifying xlPasteValues in the argument.

Exercise 2.14. Suppose that the current active workbook is "Chapter2.xlsm" and the workbook "USLifeTable1999-2001Male.xls" is open. Write a sub procedure named CopyTable to copy the mortality table (in range "A25:B134") in the second workbook to the first worksheet of the active workbook. The start cell of the destination range should be "F1."

For a complete list of properties and methods of a Range object, readers are referred to the MSDN (Microsoft Developer Network) documentation at https://msdn.microsoft.com/en-us/library/office/ff838238.aspx.

2.6   WorksheetFunction Object

The WorksheetFunction object is a container for Excel worksheet functions that can be called from VBA code. Excel 2013 contains more than 400 worksheet functions. Table 2.14 shows some of these functions.

You are familiar with using the worksheet functions in a cell formula. In this section, we introduce how to call the worksheet functions in VBA code.

Let us consider the following sub procedure:

```
1 Sub WorksheetFunctionDemo1()
```

Table 2.14: Some Worksheet Functions

Function	Description
Ceiling	Returns number rounded up to the nearest multiple of significance
Count	Counts the number of cells that contain numbers and counts numbers within the list of arguments
CountIf	Counts the number of cells within a range that meet the given criteria
Floor	Rounds number down to the nearest multiple of significance
Ln	Returns the natural logarithm of a number
Lookup	Returns a value either from a one-row or one-column range or from an array
Max	Returns the largest value in a set of values
Min	Returns the smallest number in a set of values
NormDist	Returns the normal distribution for the specified mean and standard deviation
NormInv	Returns the inverse of the normal cumulative distribution for the specified mean and standard deviation
Pi	Returns the number 3.14159265358979
Power	Returns the result of a number raised to a power
RandBetween	Returns a random integer number between two specified numbers

```
2    Range("A1").Value = WorksheetFunction.RandBetween
         (0, 100) / 100
3    Range("A2").Value = WorksheetFunction.RandBetween
         (0, 100) / 100
4    Range("A3").Value = WorksheetFunction.RandBetween
         (0, 100) / 100
5    Range("A4").Value = WorksheetFunction.RandBetween
         (0, 100) / 100
6    Range("A5").Value = WorksheetFunction.RandBetween
         (0, 100) / 100
7    Debug.Print WorksheetFunction.CountIf(Range("A1:
         A5"), ">0.5")
8    Debug.Print WorksheetFunction.Max(Range("A1:A5"))
9    Debug.Print WorksheetFunction.Min(Range("A1:A5"))
10 End Sub
```

The above sub procedure first generates five random numbers and puts them into the first five cells of Column A. Then it displays the number of values that are larger than 0.5, the maximum of the five values, and the minimum of the five values. Executing the above code produces the following output:

```
1  3
2  1
3  0
```

If you execute the above code, you may get different results, as the function RandBetween returns a random integer between two numbers you specified.

Exercise 2.15. Write a sub procedure named RandBetweenDemo to generate a random number between 0 and 1 that has four decimal places. Display the random number in the Immediate window.

In the above example, we apply worksheet functions to values in a range. We can also apply worksheet functions to numbers directly:

```
1  Sub  WorksheetFunctionDemo2()
2      Debug.Print  WorksheetFunction.Power(2,  10)
3      Debug.Print  WorksheetFunction.Pi
4      Debug.Print  WorksheetFunction.Ln(100)
5      Debug.Print  WorksheetFunction.Ceiling(0.51,  1)
6      Debug.Print  WorksheetFunction.Floor(0.51,  1)
7      Debug.Print  WorksheetFunction.NormInv(0.95,  0,  1)
8      Debug.Print  WorksheetFunction.NormDist(0,  0,  1,
          True)
9  End  Sub
```

Executing the above code produces the following output:

```
1  1024
2  3.14159265358979
3  4.60517018598809
4  1
5  0
6  1.64485362695147
7  0.5
```

The function NormDist(x, μ, σ, p_4) calculates the density (if p_4 is false) and probability (if p_4 is true) of a normal distribution with mean μ and standard deviation σ at x. The normal density function is defined as

$$f(x, \mu, \sigma) = \frac{1}{\sqrt{2\pi}\sigma} \exp\left(-\frac{(x-\mu)^2}{\sigma^2}\right).$$

(2.1)

If p_4 is false, the function NormDist(x, μ, σ, p_4) returns $f(x, \mu, \sigma)$. If p_4 is true, the function NormDist(x, μ, σ, p_4) returns the following integrated value:

$$\int_{-\infty}^{x} f(s, \mu, \sigma)ds.$$

The function NormIvs(x, μ, σ) returns the value of y, such that

$$x = \int_{-\infty}^{y} f(s, \mu, \sigma)ds.$$

Exercise 2.16. Write a sub procedure named WorksheetFunctionDemo3 to calculate the following quantities:

(a) The circumference of a circle that has a diameter of 2 meters. (The formula is πd, where d is the diameter.)

(b) The probability of a standard normal variable less than 1.5.

(c) The value of y, such that the probability of a standard normal variable less than y is 0.025.

Note that a standard normal variable has a mean of 0 and a standard deviation of 1.

A complete list of worksheet functions contained in the WorksheetFunction object can be found in the Object Browser of the Visual Basic Editor (see Figure 2.4). The usage of worksheet functions can be found in the MSDN document "WorksheetFunction Methods (Excel)" that is available at https://msdn.microsoft.com/en-us/library/office/dn301180.aspx.

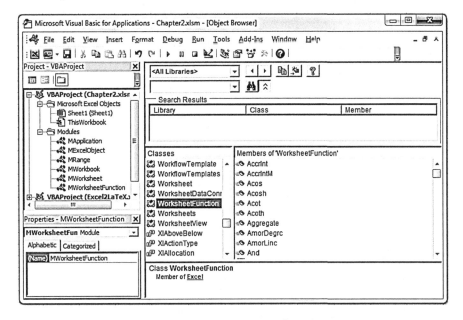

Figure 2.4: Worksheet functions in the Object Browser

2.7 Summary

In this chapter, we introduced the Excel object model, which contains hundreds of objects and thousands of properties and methods. We also introduced some commonly encountered objects, such as the `Application`, `Range`, and `Cells` objects. For an overall view of the Excel object model, readers are referred to (Roman, 2002, Chapter 15). For more information about object-oriented programming, readers are referred to (Urtis, 2015, Lesson 5).

3

Variables, Data Types, and Scopes

A variable is a named storage location in a computer's memory. Variables allow us to store values and then come back to these values at a later time. In this chapter, we introduce how to declare variables and some common data types. After studying this chapter, readers will be able to

- know the common data types supported by Visual Basic for Applications (VBA).
- know how to declare and use variables and constants.
- know how to declare and use arrays.
- know how to manipulate dates and strings.
- understand the concept of scopes.

3.1 Variable Declaration and Data Types

To declare a variable in VBA code, we first need to come up with a name for the variable. The name of a variable should satisfy the following conditions:

- The first character must be a letter.
- The following symbols cannot be used in a name: space, period (.), exclamation mark (!), and the characters @, &, $, #.
- A name cannot exceed 255 characters in length.
- A name cannot be a VBA keyword.

Exercise 3.1. Determine the legal variable names in the following list of names:

```
ws1,  1ws,  my.book,  Sub,  w b,  Tmp,  sub
```

The following sub procedure shows how to declare a variable in VBA code:

```
Sub VariableDemo1()
    Dim x As Double

    Debug.Print x
    x = 1
    Debug.Print x
    x = x + 10
    Debug.Print x
End Sub
```

The syntax of declaring a variable in VBA is shown in Line 2 of the above code. We use the VBA keyword Dim to declare variables. In this example, we declare a double-type variable named x. Then we display the value of x after some operations. Executing the above code gives the following output:

```
0
1
11
```

Table 3.1 shows a list of VBA's built-in data types. Regarding how to choose a data type for a variable, you should in general choose a data type that can hold all the data you want to store and uses the smallest number of bytes. Most data types shown in Table 3.1 are easy to understand except for Object and Variant.

We declare an Object variable using the Dim statement as above. However, we need to use the Set statement to assign an object to an object variable. The following sub procedure shows how to declare Object variables and assign objects to the variables:

```
Sub VariableDemo3()
    Dim wb As Object
    Dim ws As Object

    Set wb = ThisWorkbook
    Debug.Print wb.Name

    Set ws = ThisWorkbook.ActiveSheet
    Debug.Print ws.Name
End Sub
```

Table 3.1: List of Built-In Data Types in VBA

Data Type	Bytes Used	Range of Values
Boolean	2	True or False
Byte	1	0 to $2^8 - 1$
Integer	2	-2^{15} to $2^{15} - 1$
Long	4	-2^{31} to $2^{31} - 1$
Single	4	-3.4E+38 to -1.4E-45 for negative values, 1.4E-45 to 3.4E+38 for positive value
Double	8	-1.79E+308 to -4.94E-324 for negative values, 4.94E-324 to 1.79E+308 for positive values
Currency	8	-922,337,203,685,477.5808 to 922,337,203,685,477.5807
Date	8	1/1/100 to 12/31/9999
String	1 per character	Varies
Object	4	Any defined object
Variant	Varies	Varies

In the above code, we declare two Object variables and assign a Workbook object and a Sheet object to the variables, respectively. Executing the above code produces the following output:

```
1 Chapter1.xlsm
2 Sheet1
```

Exercise 3.2. What is wrong with the following VBA code?

```
1 Sub VariableDemo4()
2     Dim wb As Workbook
3
4     wb = Workbooks(1)
5     Debug.Print wb.Name
6 End Sub
```

Exercise 3.3. What is wrong with following code?

```
1 Sub VariableDemo5()
```

```
2       Dim  wb  As  Workbook
3
4       Set  wb  =  ThisWorkbook . ActiveSheet
5       Debug . Print  wb . Name
6  End  Sub
```

Exercise 3.4. Does the following code contain any errors?

```
1  Sub  VariableDemo6 ()
2       Dim  myObj  As  Object
3
4       Set  myObj  =  ThisWorkbook
5       Debug . Print  myObj . Name
6
7       Set  myObj  =  ThisWorkbook . ActiveSheet
8       Debug . Print  myObj . Name
9  End  Sub
```

Exercise 3.5. Write a sub procedure named WorkbookDemo6 to create a new workbook, save the new workbook to a file named "MyWorkbook.xlsx," and then close the new workbook.

Exercise 3.6. Write a sub procedure named WorksheetDemo6 to create a new worksheet and rename the new worksheet to "SheetB."

A Variant variable can contain any kind of data except fixed-length String data. The following example illustrates how to use a Variant variable:

```
1  Sub  VariableDemo7 ()
2       Dim  x  As  Variant
3
4       x  =  1
5       Debug . Print  x
6       x  =  3.54
7       Debug . Print  x
8       Set  x  =  ThisWorkbook
9       Debug . Print  x . Name
10 End  Sub
```

Executing the above code produces the following output:

```
1   1
2   3.54
3 Chapter1.xlsm
```

From the output, we see that a `Variant` variable can be used to store different types of data without modifying the data.

VBA is not a strictly typed programming language, that is, we can use a
variable without explicitly declaring it. In the following example, we assign
values to two variables and use the function `VarType` to get the data type of
the variables.

```
1 Sub VariableDemo2()
2     x = 1
3     Debug.Print VarType(x)
4     Debug.Print vbInteger
5
6     y = 3.14
7     Debug.Print VarType(y)
8     Debug.Print vbDouble
9 End Sub
```

Executing the above code produces the following output:

```
1   2
2   2
3   5
4   5
```

From the output, we see that VBA automatically treated x as an `Integer`
variable and y as a `Double` variable. In the above code, vbInteger and
vbDouble are VBA constants that denote the data types.

It is not a best practice to use a variable without explicitly declaring
it. To force explicit variable declaration, we can put `Option Explicit` in
the declaration section of a module (see Section 1.2). For example, if we
put `Option Explicit` at the beginning of the module and run the sub
procedure `VariableDemo2`, we see the compile error shown in Figure 3.1.

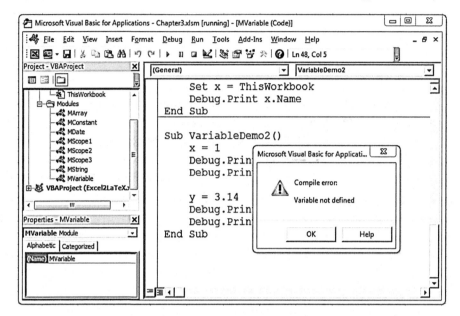

Figure 3.1: An illustration of an explicit variable declaration

3.2 Arrays

In Section 3.1, we introduced how to declare variables, which contain a single value. In this section, we introduce how to declare arrays, which can contain multiple values. In particular, we will introduce one-dimensional, multidimensional, and dynamic arrays.

An array is a group of variables that share a common name. We can refer to a specific variable in an array by using the array name and an index number in parentheses. The following example shows how to declare one-dimensional arrays:

```
1  Sub ArrayDemo1()
2      Dim arA(0 To 9) As Integer
3      Dim arB(9) As Integer
4      Dim arC(1 To 10) As Integer
5
6      Debug.Print UBound(arA)
7      Debug.Print LBound(arA)
8      Debug.Print UBound(arB)
9      Debug.Print LBound(arB)
10     Debug.Print UBound(arC)
```

```
11        Debug.Print  LBound(arC)
12 End  Sub
```

In the above sub procedure, we declare three arrays, each of which has ten elements. Inside the parentheses of arA, 0 means the lower index and 9 means the upper index of the array. The functions UBound and LBound are used to get the upper and lower bounds of the array index. If we execute the above code, we see the following output:

```
1  9
2  0
3  9
4  0
5  10
6  1
```

From the above output we see that arA and arB are identical in that the bounds of the two arrays are the same. However, the bounds of arC are 1 and 10.

Exercise 3.7. Let arD be an array defined in the following code:

```
1 Sub  ArrayDemo2()
2      Dim  arD(5  To  100)  As  Integer
3 End  Sub
```

What are the bounds of the array arD, and how many elements does the array contain?

From the above example we see that if you omit the lower index in an array declaration, VBA assumes that it is 0. However, you can change the default lower index to 1 by putting Option Base 1 in the declaration section of the module. Suppose that the underlying module is MArray, and we have Option Base 1 in the declaration section of the module. Then executing the following code

```
1 Sub  ArrayDemo3()
2      Dim  arE(5)  As  Integer
3
4      Debug.Print  LBound(arE)
```

```
5      Debug.Print UBound(arE)
6 End Sub
```

will produce the following output:

```
1  1
2  5
```

---◆---

Exercise 3.8. What is wrong with the following code?

```
1 Option Explicit
2 Option Base 10
```

Exercise 3.9. What is wrong with the following code?

```
1 Sub ArrayDemo4()
2     Dim arF(5 To 10) As Integer
3
4     Debug.Print LBound(arF)
5     Debug.Print UBound(arF)
6 End Sub
```

---◆---

Once we declare an array, we can access its elements by using the array name and an index number in parentheses. The following example shows how to get and set values of an array:

```
1 Sub ArrayDemo5()
2     Dim arA(1 To 5) As Double
3     arA(1) = 1.1
4     arA(3) = arA(1) + 2
5     arA(5) = arA(3) + 2
6
7     Debug.Print arA(1), arA(2), arA(3), arA(4), arA
           (5)
8 End Sub
```

Executing the above code gives the following output:

```
1  1.1          0          3.1          0          5.1
```

We can also declare a multidimensional array by specifying the bounds of every dimension. In the following example, we declare a two-dimensional array:

```
 1  Sub ArrayDemo6()
 2      Dim arM(1 To 3, 1 To 3) As Double
 3
 4      arM(1, 1) = 1
 5      arM(1, 2) = 3
 6      arM(1, 3) = 4
 7      arM(2, 1) = arM(1, 2)
 8      arM(2, 2) = 4
 9      arM(2, 3) = 5
10      arM(3, 1) = 4
11      arM(3, 2) = arM(2, 3)
12      arM(3, 3) = 6
13
14      Debug.Print arM(1, 1), arM(1, 2), arM(1, 3)
15      Debug.Print arM(2, 1), arM(2, 2), arM(2, 3)
16      Debug.Print arM(3, 1), arM(3, 2), arM(3, 3)
17  End Sub
```

The above example also shows how to access elements in a two-dimensional array. Executing the above code produces the following output:

```
 1  1            3            4
 2  3            4            5
 3  4            5            6
```

Exercise 3.10. Write a sub procedure named ArrayDemo7 that declares a $2 \times 2 \times 2$ array, assigns a value (e.g., the sum of the indices) to each element, and displays all the elements.

We can also use the functions LBound and UBound to get the bounds of a multidimensional array. However, we need to specify which dimension in the function calls. The following example shows how to get the bounds of a multidimensional array:

```
 1  Sub ArrayDemo8()
 2      Dim arM(1 To 10, 0 To 9, 1 To 10) As Double
```

```
3
4       Debug.Print LBound(arM, 1), LBound(arM, 2),
            LBound(arM, 3)
5       Debug.Print UBound(arM, 1), UBound(arM, 2),
            UBound(arM, 3)
6  End Sub
```

Executing the above code produces the following output:

```
1   1                  0                  1
2   10                 9                  10
```

VBA allows us to create dynamic arrays, which do not contain a preset number of elements. The following example shows how to declare a dynamic array and change the size of the array:

```
1  Sub ArrayDemo9()
2      Dim arV() As Integer
3
4      ReDim arV(1 To 3)
5      arV(1) = 1
6      arV(2) = 2
7      arV(3) = 3
8      Debug.Print arV(1), arV(2), arV(3)
9
10     ReDim arV(1 To 5)
11     Debug.Print arV(1), arV(2), arV(3), arV(4), arV
            (5)
12 End Sub
```

In the above code, we use the ReDim statement to change the size of the array. We can use the ReDim statement as many times as we want. Executing the above code gives the following output:

```
1   1           2           3
2   0           0           0           0           0
```

From the output, we see that the second ReDim statement destroyed all the old values contained in the array arV.

To keep the old values while using the ReDim statement, we need to use the Preserve statement after the ReDim statement, as shown in the following example:

```
1  Sub ArrayDemo10()
2      Dim arV() As Integer
```

```
 3
 4      ReDim arV(1 To 3)
 5      arV(1) = 1
 6      arV(2) = 2
 7      arV(3) = 3
 8      Debug.Print arV(1), arV(2), arV(3)
 9
10      ReDim Preserve arV(1 To 5)
11      Debug.Print arV(1), arV(2), arV(3), arV(4), arV
            (5)
12 End Sub
```

Executing the above code produces the following output:

```
1  1           2           3
2  1           2           3           0           0
```

From the output, we see that the old values are preserved after changing the size of the array.

Exercise 3.11. What is wrong with the following code?

```
 1 Sub ArrayDemo11()
 2      Dim arV() As Integer
 3
 4      ReDim arV(0 To 2)
 5      Debug.Print arV(0), arV(1), arV(2)
 6      arV(0) = 3
 7      arV(1) = 1
 8      arV(2) = 2
 9
10      ReDim arV(1 To 3)
11      Debug.Print arV(1), arV(2), arV(3)
12
13      ReDim Preserve arV(0 To 3)
14      Debug.Print arV(0), arV(1), arV(2), arV(3)
15 End Sub
```

3.3 Constants

A variable's value often changes during execution of a program. If we want to refer to a value or string that never changes during execution of the program, we should use a constant. A constant is a meaningful name that takes the place of a number or string that does not change during execution of a program.

The following example illustrates how to define constants:

```
Sub ConstantDemo1()
    Const conPi As Double = 3.1415
    Const conDaysInWeek As Integer = 7
    Const conVersion  As String = "v1.0"

    Debug.Print conPi
    Debug.Print conDaysInWeek
    Debug.Print conVersion
End Sub
```

In the above sub procedure, we define three constants using the `Const` statement: a real constant, an integer constant, and a string constant. Executing the above code produces the following output in the Immediate window:

```
 3.1415
 7
v1.0
```

Exercise 3.12. What is wrong with the following code?

```
Sub ConstantDemo2()
    Const conMonths As Single = 12
    Const conQuarters As Byte = 4
    Const conDebug As Boolean = False
    Const conBook as Workbook = Thisworkbook
    Const conCell as Range = Thiswookbook.cells(1,1)
    Const conMode = "Release"
    Const conCellsInRow As Integer = ThisWorkbook.
        ActiveSheet.Rows(1).Cells.Count
End Sub
```

We can also define constants in terms of previously defined constants, as shown in the following example:

```
1 Sub ConstantDemo3()
2     Const conPi As Double = 3.1415
3     Const conPi2 As Double = 2 * conPi
4
5     Debug.Print conPi2
6 End Sub
```

In the above example, we defined the constant conPi2 based on the constant conPi.

Exercise 3.13. What is wrong with the following code?

```
1 Sub ConstantDemo4()
2     Const conE2 As Double = conE * 2
3     Const conE As Double = 2.718
4
5     Debug.Print conE2
6 End Sub
```

Exercise 3.14. What is wrong with the following code?

```
1 Sub ConstantDemo5()
2     Const conPi As Double = 3.14
3     conPi = conPi + 0.00159
4
5     Debug.Print conPi
6 End Sub
```

Since we define the constants in a procedure, these constants can only be used in the procedure. If we want a constant to be available to many procedures, we can define the constant in the declaration section of a module (see Section 1.2). For example, we can move the above three constants to the declaration section of the module:

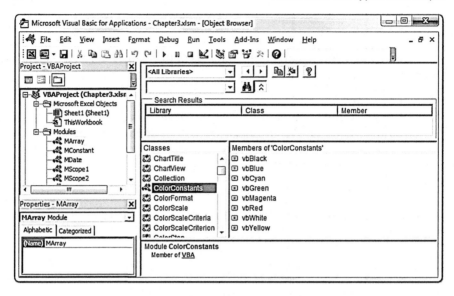

Figure 3.2: VBA's built-in color constants

```
1 Option Explicit
2
3 Const conPi As Double = 3.1415
4 Const conDaysInWeek As Integer = 7
5 Const conVersion  As String = "v1.0"
```

Then we can use these constants in all procedures in the module. In Section 3.6, we will discuss more about the scope of a constant.

We have discussed how to define our own constants. VBA also contains many built-in constants. For example, Figure 3.2 shows a list of VBA's built-in color constants, and Figure 3.3 shows a list of VBA's built-in constants for variable types.

Exercise 3.15. Write a sub procedure named ConstantDemo6 to display the values contained in the following built-in constants: vbGreen, vbYellow, vbByte, and vbError.

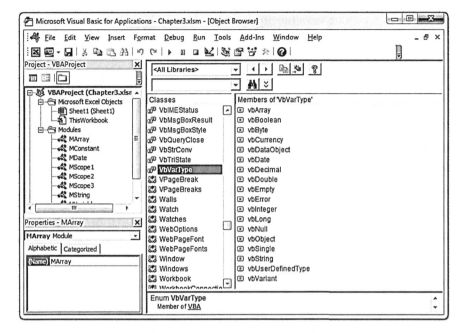

Figure 3.3: VBA's built-in constants for variable types

3.4   Strings

In many situations, we need to handle string (or text) data. VBA provides a useful set of functions for handling strings. In this section, we introduce how to create strings and use existing functions to manipulate them.

There are two types of strings in VBA: fixed-length strings and dynamic strings. The following example shows how to declare a fixed-length string:

```
Sub StringDemo1()
    Dim strFixed As String * 15

    Debug.Print Len(strFixed)
    Debug.Print strFixed
    strFixed = "I am a fixed string"
    Debug.Print Len(strFixed)
    Debug.Print strFixed
End Sub
```

In the above code, we declare a string variable `strFixed` that contains exactly 15 characters. When VBA creates this variable, it fills the variable with 15 spaces. The variable always contains 15 characters, no matter how many characters you place into it. Executing the above code produces the following output:

```
1    15
2
3    15
4  I am a fixed st
```

From the above output, we see that the function Len always returns the fixed size of the string.

Unlike fixed-length strings, dynamic strings do not have a fixed size. Creating a dynamic string is straightforward, as shown in the following example:

```
1  Sub StringDemo2()
2      Dim strDynamic As String
3
4      Debug.Print Len(strDynamic)
5      Debug.Print strDynamic
6      strDynamic = "I am a fixed string"
7      Debug.Print Len(strDynamic)
8      Debug.Print strDynamic
9  End Sub
```

When we add or remove characters from a dynamic string, VBA will take care of allocating or deallocating memory as necessary to store the text data. If we execute the above code, we see the following output:

```
1    0
2
3    19
4  I am a fixed string
```

From the output, we see that the function Len returns the actual number of characters contained in the string.

A fixed-length string cannot contain more than 65,526 characters. A dynamic string can contain as many as two billion characters. Dynamic strings are a bit slower to use because they require more processing effort from VBA. However, we need to use trim functions to remove extra spaces of a fixed-length string every time we use the string. As a result, we will use fixed-length strings only when it is necessary.

Now let us introduce VBA's built-in functions for manipulating strings.

Table 3.2: Some Commonly Used String Functions in VBA

Function	Description
Filter	Returns a zero-based array containing a subset of a String array based on specified filter criteria
Format	Returns a string formatted according to a specified format
InStr	Returns the start position of the first occurrence of one string within another
InStrRev	Returns the position of the first occurrence of one string within another, starting from the right side of the string
Join	Returns a string created by joining substrings contained in an array
LCase	Returns a string or character converted to lowercase
Left	Returns a string containing a specified number of characters from the left side of a string
Len	Returns the number of characters in a string
LSet	Returns a left-aligned string containing the specified string adjusted to the specified length
LTrim	Returns a string containing a copy of a specified string with no leading spaces
Mid	Returns a string containing a specified number of characters from a string
Replace	Returns a string where a specified substring has been replaced with another substring for a specified number of times
Right	Returns a string containing a specified number of characters from the right side of a string
RSet	Returns a right-aligned string containing the specified string adjusted to the specified length
RTrim	Returns a string containing a copy of a specified string with no trailing spaces
Space	Returns a string consisting of the specified number of spaces
Split	Returns a zero-based, one-dimensional array of substrings
StrComp	Returns −1, 0, or 1 based on the result of a string comparison
Trim	Returns a string containing a copy of a specified string with no leading or trailing spaces
UCase	Returns a string or character containing the specified string converted to uppercase

Table 3.2 gives a list of commonly used string functions provided by VBA. We first introduce how to compare two strings.

In Excel VBA, we have two methods of comparing strings: the binary mode and the text mode. In the binary mode, strings are compared based on their binary representation. In the binary mode, "A" and "a" are different. In the text mode, strings are treated case insensitively. In this mode, "A" is the same as "a." We can specify the method of comparison in a module by using the Option Compare statement in the declaration section. For example, if we want to compare strings in the text mode in a module, we can put the following line

```
1 Option Compare Text
```

in the declaration section of the module. If we do not specify the Option Compare Text statement, VBA uses Option Compare Binary.

The following example shows how to compare two strings:

```
1 Sub StringDemo3()
2       Dim strA As String
3       Dim strB As String
4
5       strA = "Hello"
6       strB = "hello"
7       Debug.Print srtA < strB
8       Debug.Print strA = strB
9       Debug.Print strA > strB
10      Debug.Print StrComp(strA, strB, vbBinaryCompare)
11      Debug.Print StrComp(strA, strB, vbTextCompare)
12 End Sub
```

Suppose that VBA uses the default comparison mode (i.e., the binary mode) in the module where the above code is saved. Then, executing the above code gives the following output:

```
1 True
2 False
3 False
4 -1
5  0
```

The above output shows that strA is less than strB in the binary mode. However, the two strings are the same if compared in the text mode. In the StrComp function, we can specify the mode of comparison.

Leading or trailing spaces are included in string comparison, as shown in the following example:

```
1 Sub StringDemo4()
2     Debug.Print StrComp("Hello", "Hello ",
          vbTextCompare)
3     Debug.Print StrComp(" Hello", "Hello",
          vbTextCompare)
4 End Sub
```

Executing the above code gives the following output:

```
1 -1
2 -1
```

The output shows that leading and trailing spaces in a string affect the result of string comparison. We use the `Trim` function to remove the leading and trailing spaces in the string as follows:

```
1 Sub StringDemo5()
2     Debug.Print StrComp("Hello", Trim("Hello ")),
          vbTextCompare)
3     Debug.Print StrComp(Trim(" Hello"), "Hello",
          vbTextCompare)
4 End Sub
```

The output of the above code is

```
1  0
2  0
```

The output shows that the two strings are the same when the spaces are removed.

———————————— ❧ ————————————

Exercise 3.16. The following function returns an array of 1000 words generated randomly:

```
1 Function RandomWords()
2     Dim arV(1 To 1000) As String
3     Dim i As Integer
4     Dim j As Integer
5     Dim k As Integer
6     Dim strW As String
7
```

```
8       Rnd (-1)
9       Randomize (1)
10      For i = 1 To 1000
11          strW = ""
12          For j = 1 To 4
13              k = Math.Round(Rnd * (Asc("e") - Asc("a")
                    + 1), 0) + Asc("a")
14              strW = strW & Chr(k)
15          Next j
16          arV(i) = strW
17      Next i
18
19      RandomWords = arV
20 End Function
```

Write a sub procedure named StringDemo6 to find out the frequency of each word. Which words are the most frequent words in the array?

———————————————————◆———————————————————

Now, we introduce how to search a string in another string. To do that, we can use the InStr function, which returns the position of the first occurrence of a string in another. Suppose that we want to search the string "car" or "Car" in the string "This car is beautiful." Then, we can use this function as follows:

```
1 Sub StringDemo7()
2      Dim strA As String
3      strA = "This car is beautiful"
4
5
6      Debug.Print InStr(1, strA, "car", vbTextCompare)
7      Debug.Print InStr(1, strA, "Car", vbTextCompare)
8      Debug.Print InStr(1, strA, "Car", vbBinaryCompare
          )
9      Debug.Print InStr(strA, "car")
10     Debug.Print InStr(strA, "Car")
11 End Sub
```

The function InStr can take four arguments. The first and the last arguments are optional. The default value of the first argument is 1, and the default value of the last argument is vbBinaryCompare. Executing the above code gives the following output:

```
1  6
2  6
3  0
4  6
5  0
```

To replace a substring in a string with another string, we can use the Replace function. The following example shows how to replace a substring in a string:

```
1 Sub StringDemo8()
2     Dim strA As String
3     Dim strB As String
4     strA = "This car is beautiful"
5
6     strB = Replace(strA, "car", "house")
7
8     Debug.Print strB
9 End Sub
```

In the above example, we use the function Replace to replace the substring "car" with "house." Executing the above code gives the following output:

```
1 This house is beautiful
```

Exercise 3.17. The function Replace has six arguments. Read the Microsoft Developer Network (MSDN) document of this function and write a sub procedure named StringDemo9 to replace only the first occurrence of "car" in the following string with "house":

> This car is beautiful. That car is big.

To extract a substring from a string, we can use the Mid function. The following example shows how to extract a substring from a string:

```
1 Sub StringDemo10()
2     Dim strA As String
3     Dim strB As String
4     Dim intA As Integer
5     strA = "This car is beautiful. That car is big."
```

```
6
7      intA = InStr(strA, ".")
8      strB = Mid(strA, 1, intA)
9
10     Debug.Print strB
11 End Sub
```

In the above example, we find the first dot in the string and then use the Mid function to extract the substring. The Mid function can take three arguments: the first argument is a string, the second one is the start position, and the third one is the length of the substring to be extracted. Executing the above code produces the following output:

```
1 This  car  is  beautiful.
```

Exercise 3.18. Write a sub procedure named StringDemo11 to replace only the second occurrence of "car" in the following string with "house":

This car is beautiful. That car is small. Your car is big.

Display the lengths of the original string and the resulting string to make sure the difference of the lengths is 2. (The word "house" has two more letters than "car.")

Exercise 3.19. Write a sub procedure named StringDemo12 to extract words enclosed by "td>" and "<" in the following string:

"<td>name</td><td>procedure</td><td>word</td>"

Display the extracted words in the Immediate window.

To split a string with a delimiter, we can use the Split function. The following example shows how to split a string:

```
1 Sub StringDemo13()
2     Dim strA As String
3     Dim arB() As String
4
5     strA = "Apple and Banana"
6     arB = Split(strA, " and ")
```

```
 7
 8        Debug.Print  LBound(arB),  UBound(arB)
 9        Debug.Print  arB(0)  &  "-"  &  arB(1)
10  End  Sub
```

In the above code, the Split function takes two arguments: the first argument is the string to be split, and the second argument is the delimiter. Executing the above code produces the following output:

```
1  0              1
2  Apple-Banana
```

Exercise 3.20. Write a sub procedure named StringDemo14 to extract the numbers in the following string into an array of double numbers

$$"1.1,0.2,3.14,4.2,0.5,4.8,1.3,6.2,1"$$

that consists of numbers separated by commas.

3.5 Dates

In many VBA programs, we need to deal with dates at one point or another. VBA provides some built-in functions for creating and manipulating date and time. Table 3.3 shows a list of VBA's built-in functions for manipulating date and time. In this section, we introduce how to use these function to manipulate date and time.

The following example shows how to get the current date and time and extract a portion from a time value:

```
1  Sub  DateDemo1()
2       Debug.Print  Now
3       Debug.Print  Date
4       Debug.Print  Year(Now)
5       Debug.Print  Month(Now)
6       Debug.Print  Day(Now)
7       Debug.Print  Hour(Now)
```

Table 3.3: VBA's Built-in Date and Time Functions

Function	Description
Date	Returns the current date
DateAdd	Add and subtract dates
DateDiff	Returns the difference in dates
DatePart	Returns a portion of a date
DateSerial	Returns a Date value representing a specified year, month, and day, with the time set to midnight
DateValue	Returns a Date value containing the date information represented by a string, with the time set to midnight
Day	Returns an Integer value from 1 through 31 representing the day of the month
Format	Formats a date according to a specified format
Hour	Returns the hour (0 to 23) of a time value
Minute	Returns the minute (0 to 59) of a time value
Month	Returns the month (1 to 12) of a date value
MonthName	Returns a string representing the month given a number from 1 to 12
Now	Returns the current date and time
WeekDay	Returns a number representing the day of the week
WeekDayName	Returns a string representing the day of the week given a number from 1 to 7
Year	Returns a four-digit year (1900 to 9999) of a date value

```
 8       Debug.Print  Minute(Now)
 9       Debug.Print  Year(Date)
10       Debug.Print  Month(Date)
11       Debug.Print  Day(Date)
12  End  Sub
```

Executing the above code produces the following output:

```
1  1/23/2016  2:28:58  PM
2  1/23/2016
3  2016
4  1
5  23
6  14
7  28
8  2016
9  1
```

10 | 23

The above example shows how to get the current date. We can also create a date variable and assign a value to it. The following sub procedure shows how to create a date value:

```
1 Sub DateDemo2()
2     Dim datA As Date
3
4     datA = #12/1/2005#
5     Debug.Print datA
6     Debug.Print Format(datA, "yyyy/mm/dd")
7     Debug.Print Format(datA, "yy/m/d")
8 End Sub
```

In the above example, we assign the value "12/1/2005" to the variable datA. We notice that the date value "12/1/2005" is surrounded by the # sign. We also use the Format function to display the date in specified formats. Executing the above code gives the following output:

```
1 12/1/2005
2 2005/12/01
3 05/12/1
```

Exercise 3.21. Write a sub procedure named DateDemo3 to get the year, month, and day of the following date value:

Jan 30, 1990

To add an interval to a date value, we can use the DateAdd function, as shown in the following example:

```
1 Sub DateDemo4()
2     Dim datA As Date
3
4     datA = #1/23/2009#
5     Debug.Print datA
6     Debug.Print DateAdd("d", 2, datA)  ' add 2 days
7     Debug.Print DateAdd("m", 3, datA)  ' add 3 months
8     Debug.Print DateAdd("ww", 1, datA) ' add 1 week
```

```
 9      Debug.Print DateAdd("q", 1, datA) ' add 1 quarter
10      Debug.Print DateAdd("yyyy", 1, datA) ' add 1 year
11 End Sub
```

Executing the above code produces the following output:

```
1 1/23/2009
2 1/25/2009
3 4/23/2009
4 1/30/2009
5 4/23/2009
6 1/23/2010
```

To find the number of time intervals between two Date values, we use the DateDiff function. The following example shows how to use this function:

```
 1 Sub DateDemo5()
 2     Dim datA As Date
 3     Dim datB As Date
 4
 5     datA = #3/20/2005#
 6     datB = #1/22/2016#
 7
 8     Debug.Print DateDiff("d", datA, datB) ' days
            between
 9     Debug.Print DateDiff("ww", datA, datB) ' calendar
            weeks between
10     Debug.Print DateDiff("w", datA, datB) ' weeks
            between
11     Debug.Print DateDiff("yyyy", datA, datB) ' years
            between
12 End Sub
```

Executing the above code gives the following output:

```
1    3960
2    565
3    565
4    11
```

Exercise 3.22. Write a sub procedure named DateDemo6 to create an array of dates starting from January 1, 2016, to May 31, 2016, at an interval of one week.

Exercise 3.23. Write a function procedure named MLK with one Integer-type argument, such that MLK(y) returns a Date value representing the Martin Luther King Day (the third Monday of January) of year y. What is the output of the following code?

```
1  Sub DateDemo7()
2      Dim i As Integer
3
4      For i = 2010 To 2020
5          Debug.Print MLK(i)
6      Next i
7  End Sub
```

Exercise 3.24. Write a function procedure named IsMLK with one Date-type argument, such that IsMLK(dat) returns True if dat is a Martin Luther King day, and False if dat is not a Martin Luther King day. What is the output of the following code?

```
1  Sub DateDemo8()
2      Debug.Print IsMLK(DateSerial(2016, 1, 18))
3      Debug.Print IsMLK(DateSerial(2015, 1, 19))
4      Debug.Print IsMLK(DateSerial(2010, 1, 20))
5  End Sub
```

3.6 Scopes

In this section, we introduce the scopes of variables, constants, and procedures.

A scope of a variable determines which modules and procedures can use the variable. The scope of a variable is determined at the time when the variable is declared. In VBA, there are three scopes available for variables: procedure-only, module-only, and public. A procedure-only variable can only be used in the procedure where the variable is declared. A module-only variable is declared in the declaration section of a module and is available to all procedures in the module. A public variable is also declared in the declaration section of a module and can be used in all procedures in all modules.

Table 3.4: Scopes of a Variable

Scope	Declaration
Procedure-only	Use a Dim or Static statement in the procedure
Module-only	Use a Dim or Private statement before the first procedure in the module
Public	Use a Public statement before the first procedure in a module

Table 3.4 lists approaches to declare a variable that has different scopes. In the following sub procedure, we declare two procedure-only variables:

```
1  Sub ScopeDemo1()
2      Dim intA As Integer
3      Static intCount As Integer
4
5      intA = intA + 1
6      intCount = intCount + 1
7
8      Debug.Print intA
9      Debug.Print intCount
10 End Sub
```

The variables intA and intCount can only be used in the procedure ScopeDemo1. The difference between the two variables is that one is declared with the Dim statement we used before, and the other is declared with the Static statement. The variable intA will be reset every time when the procedure is called. However, the variable intCount remains in existence the entire time VBA is running. A static variable is reset when any of the following occur:

- The macro generates certain run-time errors.
- VBA is halted.
- You quit Excel.
- You move the procedure to a different module.
- You reset the VBA program.

To see how to declare module-only variables, let us consider the following module named MScope2:

```
1  Private dSum As Double
2  Dim dMean As Double
3  Dim intCount As Integer
4
5  Sub ScopeDemo2()
6      dSum = dSum + WorksheetFunction.RandBetween(1,
           100)
7      intCount = intCount + 1
8
9      dMean = dSum / intCount
10 End Sub
11
12 Sub ScopeDemo3()
13     Debug.Print dSum
14     Debug.Print dMean
15     Debug.Print intCount
16 End Sub
```

The module contains three module-only variables: dSum, dMean, and intCount. Module-only variables can be declared with a Private statement or a Dim statement. The three variables are available to all procedures in this module. These variables are not available to procedures in other modules.

The module also contains two sub procedures: ScopeDemo2 and ScopeDemo3, respectively. If we run the procedure ScopeDemo2 three times and run the procedure ScopeDemo3 once, we see the following output:

```
1   90
2   30
3   3
```

Now, let us introduce how to declare public variables that are available to all procedures in all modules. Suppose that a module named MScope3 contains the following content:

```
1  ' Module MScope3
2
3  Public strMainFile As String
4
5  Sub ScopeDemo4()
6      Debug.Print strMainFile
7  End Sub
```

Then we can use the variable `strMainFile` in other modules. For example, we can write a sub procedure in the module `MScope1` as follows:

```
1  Sub scopedemo5()
2      ScopeDemo3.strMainFile = "Chapter1.xlsm"
3  End Sub
```

In the above code, we set a value to the public variable `strMainFile` by referring to the variable with its module's name.

Like variables, constants have scopes too. There are also three scopes available to a constant: procedure-only, module-only, and public. Table 3.5 presents the three scopes of a constant and how to define them. We already introduced how to define procedure-only constants in Section 3.3.

Table 3.5: Scopes of a Constant

Scope	Declaration
Procedure-only	Use a `Const` statement in the procedure
Module-only	Use a `Const` or `Private Const` statement before the first procedure in the module
Public	Use a `Public Const` statement before the first procedure in a module

---

Exercise 3.25. Let `MS1` be the name of a module containing the following content:

```
1  ' MS1
2  Public Const conPi As Doulbe = conPi2 / 2
3
4  Sub PrintPi()
5      Debug.Print conPi
6  End Sub
```

Let MS2 be the name of a module containing the following content:

```
1  ' MS2
2  Public Const conPi2 As Doulbe = conPi * 2
3
```

```
4 Sub PrintPi()
5     Debug.Print conPi2
6 End Sub
```

What is wrong with the above code?

For procedures, there are two scopes available to them: module-only and public. A module-only procedure can be called by other procedures in the same module. A public procedure can be called by procedures in other modules. Table 3.6 shows the scopes of a sub procedure and how to define them. The scopes of a function procedure can be defined similarly.

Table 3.6: Scopes of a Procedure

Scope	Declaration
Module-only	Use a `Private` Sub to define the procedure
Public	Use a `Sub` or `Public` Sub statement to define the procedure

3.7 Summary

In this chapter, we introduced how to declare variables, arrays, and constants in VBA. A variable is a named storage location in a computer's memory. An array is a group of variables that share a common name. A constant is a meaningful name that takes the place of a number or string that does not change during execution of a program. We also introduced how to manipulate strings and dates. Finally, we gave a brief introduction to the concept of scopes for variables, constants, and procedures. For more information about manipulating strings and dates in VBA, readers are referred to Getz and Gilbert (2001).

4

Operators and Control Structures

In this chapter, we introduce Visual Basic for Applications (VBA)'s built-in operators that can be used in our Excel® VBA code. We also introduce some control structures that can be used to direct the flow of a VBA program. After studying this chapter, readers will be able to

- use VBA's built-in operators.
- use the flow control statements, such as If-Then and Select-Case.
- use loops such as For-Next and Do-While.

4.1 Operators

There are four types of operators in VBA: arithmetic operators, string operators, comparison operators and logical operators.

Table 4.1: VBA's Arithmetic Operators

Operator	Description	Precedence
^	Exponentiation	1
*	Multiplication	2
/	Division	2
Mod	Modulus operator	3
+	Addition	4
−	Subtraction or negation	4

Precedence 1 has the highest precedence in the absence of brackets.

Table 4.1 gives a list of VBA's arithmetic operators. The default precedences of the operators are also listed and are applied in the absence of brackets. All the operators are apparent to most people except the modulus operator. The modulus operator finds the remainder after the division of one number by another. The following sub procedure shows how to use these operators:

```
1 Sub OperatorDemo1()
2     Debug.Print 1 + 2 * 3
3     Debug.Print (1 + 2) * 3
4     Debug.Print 2 ^ 3 * 3
5     Debug.Print 10 Mod 3
6     Debug.Print 8 - 4 / 2
7     Debug.Print (8 - 4) / 2
8 End Sub
```

Executing the above code produces the following output:

```
1 7
2 9
3 24
4 1
5 6
6 2
```

Exercise 4.1. Write a sub procedure named OperatorDemo2 to calculate the following expression:

$$-\frac{(1-2)^2}{2 \times 3^2} - \frac{1}{2}\ln(2\pi) - \ln 3.$$

Exercise 4.2. What is the output of the following code?

```
1 Sub OperatorDemo3()
2     Debug.Print 216 + 2 * 30 ^ 2
3 End Sub
```

If you multiply two large integers, VBA will raise an overflow error. For example, if you execute the following code:

```
1 Sub OperatorDemo4()
2     Debug.Print 1000 * 1000
3 End Sub
```

you will see a run-time error "Overflow." The reason is that VBA treats the two numbers as integers. Since the product of the two integers is outside the range of an integer in VBA, VBA raised the run-time error. To fix this problem, we can convert the integer to double before doing the multiplication:

```
1 Sub OperatorDemo5()
2     Debug.Print 1000# * 1000
3 End Sub
```

The symbol # after 1000 means that 1000 is a double, not an integer. We can also use the function CDbl to convert a number to a double:

```
1 Sub OperatorDemo6()
2     Debug.Print CDbl(1000) * 1000
3 End Sub
```

For a complete list of conversion functions, readers are referred to the Conversion module in the Visual Basic Editor (see Figure 4.1).

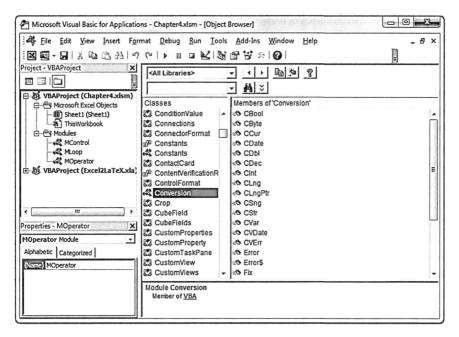

Figure 4.1: Conversion functions in the Visual Basic Editor

Exercise 4.3. Write a sub procedure named OperatorDemo7 to calculate the following quantity:

$$\sqrt{1+2015\sqrt{1+2016\sqrt{1+2017\sqrt{1+2018\sqrt{1+2019\sqrt{1+2020(2015+7)}}}}}}$$

Table 4.2: VBA's String Operator

Operator	Description
&	Concatenation operator
+	Concatenation operator

Table 4.2 shows two string operators. VBA's main string operator is the concatenation operator &, which is used to concatenate two strings. The following example shows how to use the & operator:

```
1 Sub OperatorDemo8()
2     Debug.Print "Abc" & "123"
3     Debug.Print "Abc" & CStr(123)
4     Debug.Print ThisWorkbook.Name & " is" & " " & "a
         workbook"
5 End Sub
```

In the above code, we use the function CStr to convert the number 123 to a string. Executing the above code produces the following code:

```
1 Abc123
2 Abc123
3 Chapter4.xlsm is a workbook
```

The plus sign + can also be used to concatenate two strings. If we apply + to two strings, we get a concatenated string. If we apply + to two numbers, we get the sum of the numbers. If we apply + to a number and a string, we may get an error. The following example shows the application of + in various situations:

```
1 Sub OperatorDemo9()
2     Debug.Print "abc" + "123"
3     Debug.Print "123" + 456
4     Debug.Print "123" + "456"
```

```
5        Debug.Print 123 + 456
6 End Sub
```

Executing the above code produces the following output:

```
1 abc123
2   579
3 123456
4   579
```

From the output, we see that the string "123" was converted to the number 123 automatically.

Exercise 4.4. What is wrong with the following code?

```
1 Sub OperatorDemo10()
2     Debug.Print "abc" + 123
3 End Sub
```

Exercise 4.5. What is wrong with the following code?

```
1 Sub OperatorDemo11()
2     Debug.Print "abc" & 123
3 End Sub
```

Table 4.3: VBA's Comparison Operators

Operator	Description
=	Equality
<>	Inequality
<	Less than
>	Greater than
<=	Less than or equal to
>=	Greater than or equal to
Is	Compare two object reference variables
IsNot	Compare two object reference variables

Table 4.4: VBA's Logical Operators

Operator	Description
Not	Logical opposite
And	Logical conjunction
Or	Logical disjunction or inclusion
Xor	Logical exclusion

Table 4.3 and Table 4.4 give a list of comparison operators and a list of logical operators, respectively. These operators are usually used together. The following example shows how to use these operators:

```
1  Sub OperatorDemo12()
2      Debug.Print 1 = 2
3      Debug.Print Not 1 = 2
4      Debug.Print 1 <> 2
5      Debug.Print 1 < 2
6      Debug.Print 1 > 2
7      Debug.Print 1 <> 2 Or False
8      Debug.Print 1 <> 2 And False
9      Debug.Print True Xor False
10     Debug.Print True Xor True
11     Debug.Print False Xor False
12 End Sub
```

Executing the above code produces the following output:

```
1  False
2  True
3  True
4  True
5  False
6  True
7  False
8  True
9  False
10 False
```

Exercise 4.6. What is the output of the following code?

```
1  Sub OperatorDemo13()
2      Debug.Print (1 > 2 Or 2 = 2) And (False Xor True)
3      Debug.Print (Not 1 <> 2) Xor (2 = 3 And 3 = 3)
4  End Sub
```

4.2 Flow Control

The procedures we created in previous sections start at the code's beginning and progress line by line to the end. All these procedures never divert from such top-to-bottom flow. In this section, we introduce some methods to control the flow of VBA programs. In particular, we introduce some programming constructs that allow us to skip over some statements.

Table 4.5: VBA's Programming Constructs for Flow Control

Construct	Description
GoTo statement	Jump to a particular statement
If-Then statement	Does something if a condition is true
Select Case statement	Does one of several things according to something's value

Table 4.5 shows several programming constructs that we can use to control the flow of VBA programs. A GoTo statement is a simple way to change the flow of a program. The following example shows how a GoTo statement works:

```
1  Sub GoToDemo1()
2      Debug.Print 1
3      Debug.Print 2
4      GoTo LabelA
5      Debug.Print 3
6
7  LabelA:
8      Debug.Print 4
9  End Sub
```

In the above code, we defined a label named "LabelA" in Line 7. From the code, we see that a label is just a string followed by a colon. Since we used a GoTo statement in Line 4, the code in Line 5 is skipped. Executing the above code produces the following output:

```
1  1
2  2
3  4
```

We should avoid using GoTo statements in VBA programs because using GoTo statements can make a program too complicated to understand (see Exercise 4.8). We can always use other constructs to replace GoTo statements.

Exercise 4.7. What is the output of the following code?

```vba
 1 Sub  GoToDemo2()
 2      Debug.Print  1
 3      GoTo  LabelA
 4      Debug.Print  2
 5 LabelB:
 6      Debug.Print  3
 7      GoTo  LabelC
 8      Debug.Print  4
 9
10 LabelA:
11      Debug.Print  5
12      GoTo  LabelB
13      Debug.Print  6
14
15 LabelC:
16      Debug.Print  7
17 End  Sub
```

Exercise 4.8. What is wrong with following code?

```vba
 1 Sub  GoToDemo3()
 2      Debug.Print  1
 3      GoTo  LabelA
 4      Debug.Print  2
 5 LabelB:
 6      Debug.Print  3
 7
 8 LabelA:
```

```
 9        Debug . Print 4
10        GoTo LabelB
11        Debug . Print 5
12
13 End Sub
```

An If-Then statement can also be used to change the flow of a program. The following example shows how to use an If-Then statement:

```
 1 Sub IfThenDemo1 ()
 2     If TypeOf ThisWorkbook Is Workbook Then
 3         Debug . Print "This workbook is a workbook"
 4     Else
 5         Debug . Print "This workbook is not a workbook"
 6     End If
 7
 8     If TypeOf ThisWorkbook Is Worksheet Then
 9         Debug . Print "This workbook is a Worksheet"
10     Else
11         Debug . Print "This workbook is not a Worksheet
               "
12     End If
13 End Sub
```

Exercise 4.9. The VBA function CDbl(Time) converts 24-hour time to a real number between 0 and 1, with 0 corresponding to 12:00am and 0.5 corresponding to 12:00pm. Write a sub procedure named IfThenDemo2 to display "Good Morning," "Good Afternoon," and "Good Evening" based on the time.

The If-Then statements can be nested, that is, an If-Then statement can be embedded into another one. The following example shows how to use nested If-Then statements:

```
 1 Sub IfThenDemo3 ()
 2     Dim intDay As Integer
 3     Dim intM As Integer
```

```
 4    Dim intWeekDay As Integer
 5
 6    intM = Month(Date)
 7    intDay = Day(Date)
 8    intWeekDay = Weekday(Date)
 9
10    ' In Canada, Family day is the third Monday in
         February
11    If intM = 2 Then
12        If intDay > 14 And intDay < 22 Then
13            If intWeekDay = 2 Then
14                Debug.Print "Today is Family day"
15            Else
16                Debug.Print "Today is not Family day"
17            End If
18        Else
19            Debug.Print "Today is not Family day"
20        End If
21    Else
22        Debug.Print "Today is not Family day"
23    End If
24 End Sub
```

In the above code, we have three If-Then statements. The first If-Then statement checks whether today is in February; the second statement checks whether today is in the third week of the month; and the last statement checks whether today is a Monday. The function Date returns today's date. We also use the VBA function Month, Day, and WeekDay to get the month, day, and week day from today's date, respectively.

Exercise 4.10. Labor Day is the first Monday in September. Write a sub procedure named IfThenDemo4 to check whether today is Labor Day.

Exercise 4.11. What is the output of the following code?

```
1 Sub IfThenDemo5()
2     Dim intTmp As Integer
3 LabelA:
4     intTmp = intTmp + 2
5
6     If intTmp > 10 Then
7         GoTo LabelB
8     Else
```

```
 9          GoTo LabelA
10      End If
11 LabelB:
12      Debug.Print intTmp
13 End Sub
```

If we have several cases, then it is tedious to use the If-Then statement. In such cases, we can use the Select-Case statement. The following example shows how to use the Select-Case statement in VBA:

```
 1 Sub SelectDemo1()
 2     Dim intDay As Integer
 3
 4     intDay = Day(Date)
 5
 6     Select Case intDay
 7         Case 1 To 15
 8             Debug.Print "Today is in the first half
                   of the month"
 9         Case 16 To 31
10             Debug.Print "Today is in the second half
                   of the month"
11     End Select
12 End Sub
```

The above sub procedure displays "Today is in the first half of the month" if the day is between 1 and 15 (inclusive), or displays "Today is in the second half of the month" if the day is between 16 and 31 (inclusive). We can also use a single number in a case, as shown in the following example:

```
 1 Sub SelectDemo2()
 2     Dim intWeekDay As Integer
 3
 4     intWeekDay = Weekday(Date)
 5
 6     Select Case intWeekDay
 7         Case 1
 8             Debug.Print "Today is Sunday"
 9         Case 2 To 6
10             Debug.Print "Today is a WeekDay"
11         Case 7
12             Debug.Print "Today is Saturday"
```

```
13|     End Select
14| End Sub
```

We can also use strings in cases:

```
 1| Sub SelectDemo3()
 2|     Dim strM As String
 3|
 4|     strM = Format(Date, "mmm")
 5|     Select Case strM
 6|         Case "Jan"
 7|             Debug.Print "January"
 8|         Case "Feb"
 9|             Debug.Print "February"
10|         Case Else
11|             Debug.Print "Not January or February"
12|     End Select
13| End Sub
```

In the above code, we use the function Format to convert a date into a three-letter month symbol. Then we display the full name of the month. Since there are twelve months, we did not write down all the cases. Instead, we used the Case Else to include all other cases.

Exercise 4.12. The VBA function VarType returns the data type (e.g., vbDouble, vbCurrency, vbString, etc; see VarType in the Object Browser for a complete list) of a variable. Write a sub procedure named SelectDemo4 to display the data type of the value of Cell "A1" in the active sheet.

4.3 Loops

In this section, we introduce VBA's programming constructs that can be used to execute a series of statements multiple times. These constructs are referred to as loops.

Table 4.6 shows three types of loops supported by the VBA programming language: the For-Next loop, the Do-While loop, and the Do-Until loop.

Table 4.6: VBA's Loops

Construct	Description
For-Next loop	Executes some statements for a specified number of times
Do-While loop	Executes some statements as long as something remains true
Do-Until loop	Executes some statements until something becomes true

Let us first introduce the For-Next loop, which can be used to execute a series of statements for a specified number of times.

The following example shows how to use the For-Next loop:

```
1  Sub ForNextDemo1()
2      Dim dSum As Double
3      Dim n As Integer
4
5      For n = 1 To 100
6          dSum = dSum + n
7      Next n
8
9      Debug.Print dSum
10 End Sub
```

In the above code, we use a For-Next loop to calculate the sum of integers from 1 to 100. Executing the code gives the following output:

```
1  5050
```

---◔---

Exercise 4.13. Write a sub procedure named ForNextDemo2 to calculate the sum of integers from 1 to 10,000 and display the sum in the Immediate window every 1000 steps.

Exercise 4.14. Write a sub procedure named ForNextDemo3 to calculate the following sum:

$$\sum_{i=10}^{1000} \left(i^2 + 3i^3 \right).$$

In a For-Next loop, we can also specify the step size by adding a Step statement after the For statement. The following example shows how to specify a step size:

```
 1 Sub  ForNextDemo4 ()
 2      Dim  vA (1  To  100)  As  Double
 3      Dim  n  As  Integer
 4
 5      For  n  =  LBound (vA)  To  UBound (vA)  Step  2
 6          vA (n)  =  1
 7      Next  n
 8
 9      Debug . Print  vA (1) ;  vA (2) ;  vA (3) ;  vA (4) ;  vA (5)
10 End  Sub
```

In the above code, we declare an one-dimensional array with 100 elements and use a For-Next loop to assign 1 to the elements with a step size of 2. Then, we display the first five elements of the array. Executing the code produces the following output:

```
 1   1    0    1    0    1
```

Note that the semicolon ";" after the Print function means printing immediately after one another.

Exercise 4.15. Write a sub procedure named ForNextDemo5 to create the following one-dimensional array

$$(0, 1, 0, 0, 1, 0, 0, 1, 0, \ldots, 0, 1, 0),$$

which contains 99 elements.

Exercise 4.16. Write a sub procedure named ForNextDemo6 to create the following one-dimensional array,

$$(2, 1, 3, 2, 1, 3, 2, 1, 3, \ldots, 2, 1, 3),$$

which contains 30 elements.

Exercise 4.17. Write a sub procedure named ForNextDemo7 to create the following 100×100 array,

$$\begin{pmatrix} 0 & 1 & 0 & \cdots & 0 \\ 1 & 0 & 1 & \cdots & 0 \\ 0 & 1 & 0 & \cdots & 0 \\ \vdots & \vdots & \vdots & \ddots & \vdots \\ 0 & 0 & 0 & \cdots & 0 \end{pmatrix},$$

which has ones in its subdiagonal elements and zeros in other elements.

———————————————————•———————————————————

Like the If-Then statement, the For-Next statement can be nested. Suppose that we want to calculate the following sum:

$$\sum_{i=1}^{10} \sum_{j=1}^{10} (i+j)(j+1).$$

Then, we can use nested For-Next statements as follows:

```
 1 Sub ForNextDemo8()
 2     Dim dSum As Double
 3     Dim i As Integer
 4     Dim j As Integer
 5
 6     For i = 1 To 10
 7         For j = 1 To 10
 8             dSum = dSum + (i + j) * (j + 1)
 9         Next j
10     Next i
11
12     Debug.Print dSum
13 End Sub
```

———————————————————•———————————————————

Exercise 4.18. Write a sub procedure named ForNextDemo9 to create the following 100×100 array:

$$\begin{pmatrix} 0 & 1 & 2 & 3 & \cdots & 99 \\ 1 & 2 & 3 & 4 & \cdots & 100 \\ 2 & 3 & 4 & 5 & \cdots & 101 \\ \vdots & \vdots & \vdots & \vdots & \ddots & \vdots \\ 99 & 100 & 101 & 102 & \cdots & 198 \end{pmatrix}.$$

Then put the array into the range "A1:CV100" of the active worksheet.

Exercise 4.19. Write a sub procedure named ForNextDemo10 to calculate the following sum:

$$\sum_{i=1}^{20}\sum_{j=1}^{10}\frac{i^3}{ij+5}.$$

Exercise 4.20. What is wrong with the following code?

```
1  Sub  ForNextDemo11()
2        Dim  dSum  As  Double
3        Dim  i  As  Integer
4        Dim  j  As  Integer
5
6        For  i  =  1  To  20
7            For  j  =  1  To  10
8                dSum  =  dSum  +  i  *  j
9            Next  i
10       Next  j
11
12       Debug.Print  dSum
13  End  Sub
```

We can terminate a For-Next loop by using an Exit For statement inside the loop. If we want to find the smallest integer N, such that

$$\sum_{i=1}^{N}i^2 > 1000,$$

we can use the following sub procedure:

```
1  Sub  ForNextDemo12()
2        Dim  dSum  As  Double
3        Dim  n  As  Integer
4
5        For  n  =  1  To  1000
6            dSum  =  dSum  +  n  ^  2
7            If  dSum  >  1000  Then
8                Exit  For
9            End  If
10       Next  n
11
```

```
12      Debug.Print  n
13      Debug.Print  dSum
14 End  Sub
```

Executing the above sub procedure produces the following output:

```
1   14
2   1015
```

Exercise 4.21. The IsEmpty function can be used to check whether a cell is empty or not. Write a sub procedure named ForNextDemo13 to display the row index of the first empty cell in the first column of the active worksheet.

Now, let us introduce the Do-While loop, which relies on a condition to terminate the loop. The Do-While loop is suitable for situations when you do not know how many iterations you want to execute the code. For example, we can use a Do-While loop to find the smallest integer N, such that

$$\sum_{i=1}^{N} i^2 > 1000.$$

We used a For-Next loop with an Exit statement to find such an integer. However, we specified the maximum number of iterations in the For-Next loop. If we use a Do-While loop, we do not need to specify the maximum number of iterations. The following sub procedure shows how a Do-While loop works:

```
1 Sub  DoWhileDemo1()
2      Dim  dSum  As  Double
3      Dim  n  As  Integer
4
5      Do  While  dSum  <=  1000
6           n  =  n  +  1
7           dSum  =  dSum  +  n  ^  2
8      Loop
9
10     Debug.Print  n
11     Debug.Print  dSum
12 End  Sub
```

In the above code, we do not specify a maximum number of iterations. However, we increase the value of n by 1 at each iteration.

In a Do-While loop, we need to make sure the condition will become false after a fixed number of iterations. Otherwise, the loop becomes an infinite loop and will not stop (see Exercise 4.22).

Exercise 4.22. What is wrong with the following code?

```
1  Sub DoWhileDemo2()
2      Dim dSum As Double
3      Dim n As Integer
4
5      Do While dSum <= 1000
6          dSum = dSum + n ^ 2
7      Loop
8
9      Debug.Print n
10     Debug.Print dSum
11 End Sub
```

Exercise 4.23. Write a sub procedure named DoWhileDemo3 that uses a Do-While loop to calculate the following sum:

$$\sum_{i=1}^{100} i.$$

The Do-Until statement is similar to the Do-While statement in terms of syntax. However, the two loop statements are different in how they handle the testing condition. A program continues to execute statements inside a Do-While loop while the condition remains true. A program continues to execute statements in a Do-Until loop until the condition is true.

Let us consider the example of finding the smallest integer N, such that

$$\sum_{i=1}^{N} i^2 > 1000.$$

We can find the integer using a Do-Until loop as follows:

```vba
1  Sub DoUntilDemo1()
2      Dim dSum As Double
3      Dim n As Integer
4
5      Do Until dSum > 1000
6          n = n + 1
7          dSum = dSum + n ^ 2
8      Loop
9
10     Debug.Print n
11     Debug.Print dSum
12 End Sub
```

Executing the above code produces the following output:

```
1  14
2  1015
```

From the above example, we see that the Do-Until statement and the Do-While statement are exchangeable by just changing the condition statement.

4.4 Summary

In this chapter, we introduced VBA's built-in operators and some control structures. VBA provides four types of operators: arithmetic operators, string operators, comparison operators, and logical operators. The control structures in VBA include If-Then, Select-Case, and GoTo. VBA loops include For-Next, Do-While, and Do-Until.

5

Functions, Events, and File IO

As we mentioned before, a module contains two types of procedures: sub procedures and function procedures. We already introduced sub procedures. In this chapter, we introduce how to create function procedures. We also introduce event handlers and methods for reading and writing files. After studying this chapter, readers will be able to

- create functions.
- create event handlers.
- read and write text files.

5.1 User-Defined Functions

The difference between a sub procedure and a function procedure is that the former does not return a result to the caller, but the later returns a result to the caller. In the following example, we create a function named MySum with two double-type arguments:

```
1 Function MySum(x As Double, y As Double)
2     MySum = x + y
3 End Function
```

The function MySum can be called by other functions or procedures. For example, we can use the above function as follows:

```
1 Sub FunctionDemo1()
2     Debug.Print MySum(1, 2)
3     Debug.Print MySum(MySum(2, 3), 4)
4 End Sub
```

Executing the above code produces the following output in the Immediate window:

1	3
2	9

The function can also be used in cell formulas. For example, we can use the function in Cell "C1" by typing "=MySum(A1,B1)" in Cell "C1." Then, the value of Cell "C1" will be the sum of the two numbers in Cells "A1" and "B1."

Exercise 5.1. Write a function procedure named MySum2 with one Long-type argument, such that MySum2(n) returns the following value:

$$\sum_{i=1}^{n} i^2.$$

What is the output of the following code?

```
1 Sub FunctionDemo2 ()
2     Debug.Print MySum2 (10)
3     Debug.Print MySum2 (100)
4 End Sub
```

Exercise 5.2. Write a function procedure named DNorm with three Double-type arguments, such that DNorm(x, μ, σ) calculates the following value

$$f(x, \mu, \sigma) = \frac{1}{\sqrt{2\pi}\sigma} \exp\left(-\frac{(x-\mu)^2}{2\sigma^2}\right),$$

where exp($\cdot$) is the exponential function, i.e., exp(y) = e^y. What is the output of the following code?

```
1 Sub FunctionDemo3 ()
2     Debug.Print DNorm (0, 0, 1)
3     Debug.Print DNorm (0, 2, 3)
4     Debug.Print WorksheetFunction.NormDist (0, 0, 1,
         False)
5     Debug.Print WorksheetFunction.NormDist (0, 2, 3,
         False)
6 End Sub
```

You can use VBA's built-in function EXP to calculate exp($\cdot$).

In the above examples, the arguments are single numbers. In fact, we can pass arrays to a function procedure. The following function takes an array-type argument:

```
 1 Function MySum3(x() As Double)
 2     Dim dSum As Double
 3     Dim i As Integer
 4
 5     For i = LBound(x) To UBound(x)
 6         dSum = dSum + x(i)
 7     Next i
 8
 9     MySum3 = dSum
10 End Function
```

The above function calculates the sum of all elements of the input array. We can test this function using the following sub procedure:

```
 1 Sub FunctionDemo4()
 2     Dim arA(1 To 100) As Double
 3     Dim i As Integer
 4
 5     For i = 1 To 100
 6         arA(i) = i
 7     Next i
 8
 9     Debug.Print MySum3(arA)
10 End Sub
```

Executing the above code gives the following output:

```
 1    5050
```

Exercise 5.3. The sample mean and the sample standard deviation of a vector of numbers can be calculated incrementally as follows. Let $y_1, y_2, \ldots, y_n$ be n numbers. Let $\{M_{1,i} : i = 1, 2, \ldots, n\}$ and $\{M_{2,i} : i = 1, 2, \ldots, n\}$ be two sequences defined as follows:

$$M_{1,1} = y_1, \quad M_{1,k} = M_{1,k-1} + \frac{y_k - M_{1,k-1}}{k}, \quad k = 2, 3, \ldots, n,$$

and

$$M_{2,1} = 0, \quad M_{2,k} = M_{2,k-1} + \frac{k-1}{k}(y_k - M_{1,k-1})^2, \quad k = 2,3,\ldots,n.$$

Then,

$$\bar{y} = M_{1,n}$$

and

$$s_y = \sqrt{\frac{M_{2,n}}{n-1}}, \quad n > 1,$$

where $\bar{y}$ and s_y denote the sample mean and the sample standard deviation of $y_1, y_2, \ldots, y_n$.

Write two function procedures named IncMean and IncStd with one array-type argument, such that IncMean(y) and IncStd(y) calculate the sample mean and the sample standard deviation of the array y incrementally, respectively. What is the output of the following code?

```
1  Sub FunctionDemo5()
2      Dim vA(1 To 100) As Double
3      Dim i As Integer
4      For i = 1 To 100
5          vA(i) = Rnd() * 10000
6      Next i
7
8      Debug.Print IncMean(vA)
9      Debug.Print IncStd(vA)
10     Debug.Print WorksheetFunction.Average(vA)
11     Debug.Print WorksheetFunction.StDev(vA)
12 End Sub
```

Function procedures can be recursive, that is, a function procedure can call itself. For example, we consider the Fibonacci numbers $F_0, F_1, \ldots$ defined as follows:

$$F_n = F_{n-1} + F_{n-2}, \quad n \geq 2, \quad F_0 = 0, F_1 = 1. \tag{5.1}$$

We can write a nested function procedure to calculate the Fibonacci number at an arbitrary location n as follows:

```
1 Function Fibonacci(n As Long)
2     If n = 0 Then
3         Fibonacci = 0
4     ElseIf n = 1 Then
5         Fibonacci = 1
6     Else
7         Fibonacci = Fibonacci(n - 1) + Fibonacci(n -
             2)
8     End If
9 End Function
```

The above function `Fibonacci(n)` returns the Fibonacci number F_n. To test the above function, we run the following code:

```
1 Sub FunctionDemo6()
2     Debug.Print Fibonacci(10)
3     Debug.Print Fibonacci(20)
4 End Sub
```

which produces the following output:

```
1 55
2 6765
```

Exercise 5.4. Write a function procedure named `Factorial` with one Long-type argument, such that `Factorial(n)` calculates the following value

$$n! = 1 \times 2 \times \cdots n = \prod_{i=1}^{n} i.$$

We define $0! = 1$. What is the output of following code?

```
1 Sub FunctionDemo7()
2     Debug.Print Factorial(0)
3     Debug.Print Factorial(10)
4     Debug.Print Factorial(20)
5 End Sub
```

Exercise 5.5. Recursive functions are usually not efficient. We can almost always substitute a loop for recursion. Write a nonrecursive function procedure named `Fibonacci2` with one Long-type argument, such that

Fibonacci2(n) calculates the Fibonacci number F_n defined in Equation (5.1).

The function we discussed so far returns a single value. In fact, a function can also return an array. The following code shows a function that returns an array.

```
 1 Function RandNumbers(n As  Integer)
 2     Dim arA() As Double
 3     Dim i As Integer
 4
 5     ReDim arA(1 To n)
 6     Rnd (-1)
 7     Randomize (0)
 8     For i = 1 To n
 9         arA(i) = Rnd()
10     Next i
11
12     RandNumbers = arA
13 End Function
```

In the above code, the function RandNumbers has one Integer-type argument and returns an array of random numbers. We use the VBA functions Rnd and Randomize to generate random numbers. The function Randomize is used to fix the seed of the random number generator so that we can repeat the random numbers every time we call the function RandNumbers. To test this function, we use the following sub procedure:

```
 1 Sub  FunctionDemo8()
 2     Dim arA As  Variant
 3
 4     arA = RandNumbers(10)
 5     Debug.Print  LBound(arA), UBound(arA)
 6     Debug.Print  arA(1), arA(2), arA(3)
 7 End Sub
```

Executing the above sub procedure, we see the following output:

```
 1 1              10
 2 0.332842886447906            0.320361137390137
     0.087237536907196
```

Exercise 5.6. What is the difference of the following two functions?

```
1  Function RandNumbers(n As Integer)
2      Dim arA() As Double
3      Dim i As Integer
4
5      ReDim arA(1 To n)
6      Rnd (-1)
7      Randomize (0)
8      For i = 1 To n
9          arA(i) = Rnd()
10     Next i
11
12     RandNumbers = arA
13 End Function
14
15 Function RandNumbers2(n As Integer)
16     Dim arA() As Double
17     Dim i As Integer
18
19     ReDim arA(1 To n)
20     Randomize (0)
21     For i = 1 To n
22         arA(i) = Rnd()
23     Next i
24
25     RandNumbers2 = arA
26 End Function
```

5.2 Events

A Visual Basic for Applications (VBA) sub procedure can be executed automatically when a particular event occurs. In this section, we introduce some Excel® event types and where to place event-handler VBA code.

Table 5.1 gives a list of workbook events. To add a workbook event-handler, we need to double click ThisWorkbook in the VBAProject window

Table 5.1: Workbook Events

Event	Description
Activate	The workbook is activated
BeforeClose	The workbook is closed
BeforeSave	The workbook is saved
Deactivate	The workbook is deactivated
NewSheet	A new sheet is added to the workbook
Open	The workbook is opened
SheetActivate	A sheet in the workbook is activated
SheetChange	A cell in the workbook is changed

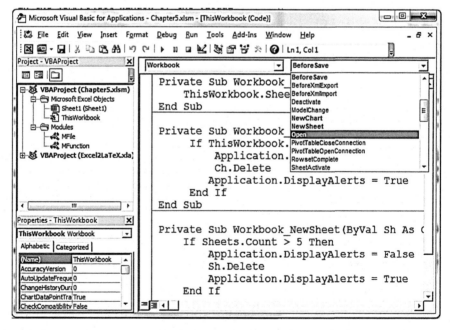

Figure 5.1: Workbook events in the code window

of the Visual Basic Editor, select Workbook in the left drop-down list of the code window, and then select the event type in the right drop-down list of the code window (see Figure 5.1). To add an event-handler to the NewSheet event, for example, we select the NewSheet event from the drop-down list. The following code is an example of the event-handler for the NewSheet event:

```
1 Private Sub Workbook_NewSheet(ByVal Sh As Object)
2     MsgBox "A new sheet named " & " " & Sh.Name & "
          is added"
3 End Sub
```

The event-handler is a sub procedure with one argument. The argument is the new sheet added to the workbook. If we add a new sheet to the workbook, we will see a message box that shows the sheet's name.

Exercise 5.7. Write an event-handler for the NewSheet event, such that a new sheet will be deleted without alert as soon as it is created.

Exercise 5.8. Add a piece of VBA code to a workbook, such that the total number of sheets (including charts and worksheets) will not exceed 5.

Exercise 5.9. Write a piece of VBA code for your workbook, such that whenever a workbook is saved, a time stamp is written to the cell "D1" of the sheet "Sheet1" of your workbook.

Table 5.2: Worksheet Events

Event	Description
Activate	The worksheet is activated
Change	A cell in the worksheet is changed
Deactivate	The worksheet is deactivated
SelectionChange	The selection is changed

From the above example, we see that event-handlers for workbook events have to sit in the workbook object. Similarly, event-handlers for worksheet events have to sit in a worksheet object. Table 5.2 shows some worksheet events.

Suppose that the cell "C3" of the worksheet "Sheet1" is a positive number that is an input for some program. We need to make sure the number in the cell is positive. We can add an event-handler for the Change event. To do that, we double click the worksheet named "Sheet1" and select the Change event from the right drop-down list in the code window. Then, we add the following code:

```
1  Private Sub Worksheet_Change(ByVal Target As Range)
2      If Target.Parent.Name = "Sheet1" And Target.
       Address = "$C$3" Then
3          If Target.Value <= 0 Then
4              MsgBox "Invalid number"
5          End If
6      End If
7  End Sub
```

The above code is executed when the name of the sheet is "Sheet1" and the address of the changed cell is $C$3. If we input −1 in the cell C3 of the sheet "Sheet1," we will see a message box.

———————————————————⚬———————————————————

Exercise 5.10. Add a piece of code to a worksheet, such that whenever the worksheet is activated, a message box with the worksheet's name is displayed.

———————————————————⚬———————————————————

5.3 File IO

In this section, we introduce how to use VBA to read and write text files.

Let us introduce how to write text files using VBA code. Suppose that the range "A1:B110" of the sheet "Sheet1" contains a mortality table. To write the mortality table to a Comma Separated Values (CSV) file, we can use the following VBA code:

```
1  Sub FileDemo1()
2      Dim strFileName As String
3      Dim ws As Worksheet
4      Dim i As Integer
```

```
5
6       Set ws = ThisWorkbook.Sheets("Sheet1")
7       strFileName = "MaleTable.csv"
8       Open strFileName For Output As #1
9       For i = 1 To 110
10          Write #1, ws.Cells(i, 1), ws.Cells(i, 2)
11      Next i
12      Close #1
13
14  End Sub
```

In the above code, we open a file for output in Line 8. The file object is denoted by #1, as it is the first file we opened. In Lines 9–11, we write 110 lines to a CSV file named "MaleTable.csv." In Line 12, we close the CSV file. Executing the above code will produce a CSV file containing the numbers in the range "A1:B110."

If we open the file "MaleTable.csv" created by the above VBA code, we see that all numbers are surrounded by double quotes. To write a CSV file without double quotes, we can use the following VBA code:

```
1   Sub FileDemo2()
2       Dim strFileName As String
3       Dim ws As Worksheet
4       Dim i As Integer
5
6       Set ws = ThisWorkbook.Sheets("Sheet1")
7       strFileName = "MaleTable.csv"
8       Open strFileName For Output As #1
9       For i = 1 To 110
10          Print #1, ws.Cells(i, 1) & "," & ws.Cells(i,
                2)
11      Next i
12      Close #1
13
14  End Sub
```

The difference between the sub procedures FileDemo2 and FileDemo1 is that we use the Print function in the former. The Print function writes a string to the file. If we write the CSV file using FileDemo2, we will not see the double quotes around the numbers.

In the sub procedure FileDemo2, we concatenate the values using the & operator and write the resulting string as a new line to the file. In fact, we do not have to concatenate the individual components before writing a line. We can write each component of a line as follows:

```
1  Sub FileDemo3()
2      Dim strFileName As String
3      Dim ws As Worksheet
4      Dim i As Integer
5
6      Set ws = ThisWorkbook.Sheets("Sheet1")
7      strFileName = "MaleTable.csv"
8      Open strFileName For Output As #1
9      For i = 1 To 110
10         Print #1, ws.Cells(i, 1);
11         Print #1, ",";
12         Print #1, ws.Cells(i, 2)
13     Next i
14     Close #1
15 End Sub
```

In the above code, the semicolon ";" after the Print command tells the command not to write a new line break.

Exercise 5.11. The following VBA sub procedure

```
1  Sub GenerateMatrix()
2      Dim i As Integer
3      Dim j As Integer
4
5      Rnd (-1)
6      Randomize (0)
7      For i = 1 To 100
8          For j = 1 To 10
9              Cells(i, j).Value = Rnd()
10         Next j
11     Next i
12 End Sub
```

writes a 100×10 matrix to the range "A1:J100." Write a sub procedure named FileDemo4 that writes the matrix to a CSV file named matrix.csv, such that each line of the file contains a row of the matrix.

Now, let us introduce how to read text files in VBA. The following example shows how to read a CSV file:

```
1 Sub FileDemo5()
2     Dim strFile As String
3     Dim strV() As String
4     Dim j As Integer
5     Dim n As Integer
6     strFile = ThisWorkbook.Path & ":matrix.csv"
7     ' strFile is a full file name with path in a Mac
         computer
8
9     n = 1
10    Open strFile For Input As #1
11    Do Until EOF(1)
12        Line Input #1, Line
13        strV = Split(Line, ",")
14        For j = 0 To UBound(strV)
15            Cells(n, j + 1).Value = strV(j)
16        Next j
17        n = n + 1
18    Loop
19    Close #1
20 End Sub
```

Let us examine the above code line by line. In Lines 2–5, we declare variables. In Line 6, we create the full file name by combining the path of the workbook and "matrix.csv," which is the intended file name. By doing that, we save the file "matrix.csv" into the same directory as this workbook. In Line 10 of the above code, we open the file for input. In Lines 11–18, we use a loop to read the CSV file line by line. The function EOF returns a boolean value indicating whether the end of the file is reached. Inside the loop, we split the line by commas and write the individual component to cells of the active worksheet.

———————————————————◆———————————————————

Exercise 5.12. Suppose that the file words.txt contains the following two lines:

```
1 Microsoft Excel is an extremely powerful tool
2 You can use to manipulate and present data
```

Write a sub procedure named FileDemo6 to do the following: read words from the file `words.txt`, write the words in the first line to the first row of your active worksheet, and write the words in the second line to the second row of your active worksheet. Each cell contains one word.

Exercise 5.13. The file `keywords.html` is an HTML file that contains VBA keywords, which are enclosed by the strings "(v=vs.90).aspx">" and "<."

(a) Write a function procedure named `ExtractKeywords` with one String-type argument, such that `ExtractKeywords(line)` returns a string that contains keywords in `line` separated by spaces.

(b) Write a sub procedure named `FileDemo7` to extract all keywords in the file `keywords.html` and write each keyword to a cell in the first column of your active worksheet.

5.4 Summary

In this chapter, we introduced how to create VBA functions, how to add a handler to an event, and how to read and write files. A function allows us to return some result to the caller of the function. An event handler enables us to execute some VBA code automatically. The file Input-Output (IO) functions allow us to import and export data from text files in a controlled way. For more information about file IO, readers are referred to (Getz and Gilbert, 2001, Chapter 12).

6

Error Handling and Debugging

There are three categories of programming errors in general: compilation errors, runtime errors, and logic errors. Compilation errors are errors that prevent your program from running. When you run your Visual Basic for Applications (VBA) code, the Visual Basic compiler compiles the VBA code into binary language. If the compiler does not understand some of your VBA code, it issues compilation errors. For example, misspelling a VBA keyword is an example of compilation errors. Runtime errors are errors that occur while your program is actually running. For example, division by zero and taking square root of a negative number are runtime errors. Logic errors are errors that prevent your VBA program from doing what you intended it to do. For example, forgetting to terminate an infinite loop is a logic error.

Compilation errors are the easiest to handle, as they can be detected by the Visual Basic Editor. Logic errors are the hardest to find and fix. In this chapter, we introduce techniques for handling runtime errors and debugging tools for finding logic errors. We will also introduce some best practices of VBA coding that can help minimize the number of errors. After studying this chapter, readers will be able to

- add error-handling code to VBA programs.
- use the debugging tools provided by VBA to debug VBA programs.
- know some best practices of VBA coding.

6.1 Error Handling

In this section, we introduce some techniques for handling runtime errors. To illustrate runtime errors, we consider the following sub procedure:

```
1  Sub ErrorDemo1()
2      Dim dVal As Double
3
```

```
4    dVal = InputBox("Input a number")
5    MsgBox "The square root of " & dVal & " is " & Sqr
        (dVal)
6  End Sub
```

The above code asks the user for a number and then shows the square root of the number in a message box. The above code is free of compilation errors. However, it contains runtime errors because calculating the square root of a negative number is illegal. For example, if we run the code and input −1 in the input box, as shown in Figure 6.1(a), then we see the runtime error shown in Figure 6.1(b).

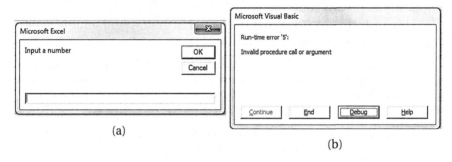

(a)

(b)

Figure 6.1: An example illustrating a runtime error

The error message shown in Figure 6.1(b) is not helpful for many people. A better way to handle the runtime error is to anticipate the error by adding error-handling code. For example, we can add error-handling code to the above sub procedure as follows:

```
1  Sub ErrorDemo2()
2    Dim dVal As Double
3
4    dVal = InputBox("Input a number")
5    If dVal < 0 Then
6        MsgBox "You must input a non-negative number."
7        Exit Sub
8    End If
9    MsgBox "The square root of " & dVal & " is " & Sqr
        (dVal)
10 End Sub
```

In Lines 5–8, we use an If-Then statement to check whether the input number is negative. If the number is negative, we display a message and use the statement Exit Sub to terminate the sub procedure.

The sub procedure `ErrorDemo2` improves the sub procedure `ErrorDemo1` by adding the error-handling code. However, it is still not perfect. For example, if we input "a" in the input box (see Figure 6.2(a)), we will see the runtime error shown in Figure 6.2(b).

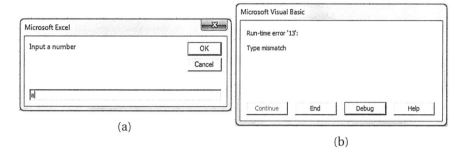

(a)

(b)

Figure 6.2: An example illustrating another runtime error

To improve the code further, we add another piece of error handling code as follows:

```
1 Sub ErrorDemo3()
2     Dim dVal As Variant
3
4     dVal = InputBox("Input a number")
5     If Not IsNumeric(dVal) Then
6         MsgBox "You must input a number"
7         Exit Sub
8     End If
9     If dVal < 0 Then
10        MsgBox "You must input a non-negative number."
11        Exit Sub
12    End If
13    MsgBox "The square root of " & dVal & " is " & Sqr
          (dVal)
14 End Sub
```

In the above code, we change the type of the variable dVal to Variant. Then we use the function IsNumeric to check whether the input is a number. If it is not a number, we display a message and exit the sub procedure.

Exercise 6.1. Consider the following function procedure:

```
1 Sub CalculateSpeed(dMiles As Double, dTime As Double)
```

```
2        CalculateSpeed = dMiles / dTime
3 End Sub
```

Does the above function contain runtime errors? If so, improve the function by adding error-handling code.

Exercise 6.2. Consider the following sub procedure:

```
1 Sub ErrorDemo4()
2    Dim dVal As Variant
3
4    dVal = InputBox("Input a number")
5    If Not IsNumeric(dVal) Then
6        MsgBox "You must input a number"
7        Exit Sub
8    End If
9    If dVal < 0 Then
10       MsgBox "You must input a non-negative number."
11       Exit Sub
12   End If
13   ActiveSheet.Cells(1, 1).Value = Sqr(dVal)
14 End Sub
```

Does the above code contain runtime errors? If so, what are the possible runtime errors?

———————————————————⌖———————————————————

Since it is hard for us to anticipate all possible runtime errors (see Exercise 6.2), how can we handle every possible runtime error? Fortunately, we can use the On Error statement to handle all errors. Table 6.1 gives some On Error statements for error-handling.

For example, we can use the On Error statement as follows:

```
1 Sub ErrorDemo5()
2    Dim dVal As Double
3    On Error GoTo BadInput
4
5    dVal = InputBox("Input a number")
6    MsgBox "The square root of " & dVal & " is " & Sqr
         (dVal)
7
8    Exit Sub
9 BadInput:
10   MsgBox "Error number " & Err.Number & " occurred."
```

Table 6.1: Usage of the On Error Statement

Statement	Description
On Error GoTo label	After executing this statement, VBA resumes execution at the line with the label
On Error GoTo 0	After executing this statement, VBA resumes its normal error-checking behavior
On Error Resume Next	After executing this statement, VBA ignores all errors and resumes execution with the next statement

```
11 End Sub
```

The above code is able to trap all errors. If an error occurs, the code will display the error number in a message box. Here we use the Err.Number to get the error number. Note that we use the Exit Sub statement right before the BadInput label. This statement is necessary because we do not want to execute the error-handling code if no errors occur.

Table 6.2: Usage of the Resume Statement

Statement	Description
Resume	After executing this statement, VBA resumes with the statement that caused the error
Resume Next	After executing this statement, VBA ignores all errors and resumes execution with the next statement
Resume label	After executing this statement, VBA resumes execution at the line with the label

In the sub procedure ErrorDemo5, we add error-handling code to display an error message and exit the sub procedure. In some cases, we want to recover from the error. In such cases, we use the Resume statement. Table 6.2 gives the usage of the Resume statement. The following example shows how to use the Resume statement:

```
1 Sub ErrorDemo6()
2     Dim dVal As Double
3     Dim intAns As Integer
4     Dim strMsg As String
5
```

```
 6 TryAgain:
 7     On Error GoTo BadInput
 8
 9     dVal = InputBox("Input a number")
10     MsgBox "The square root of " & dVal & " is " & Sqr
           (dVal)
11
12     Exit Sub
13 BadInput:
14     strMsg = "Error number " & Err.Number & " occurred
           ."
15     strMsg = strMsg & vbNewLine & "Do you want to try
           again?"
16     intAns = MsgBox(strMsg, vbYesNo + vbCritical)
17     If intAns = vbYes Then
18         Resume TryAgain
19     End If
20 End Sub
```

In Line 18 of the above code, we use the Resume TryAgain statement, where TryAgain is a label defined in Line 6. The difference between Resume and GoTo is that the Resume statement clears the error condition before continuing, while the GoTo statement does not clear error conditions.

————————————————⌖————————————————

Exercise 6.3. Consider the following code:

```
1 Sub ErrorDemo7()
2     Dim i As Integer
3
4     For i = 1 To 100
5         Cells(i, 2).Value = Sqr(Cells(i, 1).Value)
6     Next i
7 End Sub
```

The above sub procedure calculates the square root of numbers in the first column of the active worksheet and places the results into the second column. Improve the sub procedure by adding error-handling code so that the code can run through without showing error dialogs.

————————————————⌖————————————————

6.2 Debugging VBA Code

In this section, we introduce some tools that can help us identify logic errors, which are also referred to as bugs.

Common methods for debugging VBA code include examining the code, inserting `MsgBox` statements, insert `Debug.Print` statements, and using VBA's built-in debugging tools. The first method requires knowledge and experience. The second and third methods are similar. In previous sections, we have used the `Debug.Print` to display information in the Immediate window. In this section, we focus on the last method, which are much more powerful than the first three methods. In particular, we introduce how to set breakpoints in VBA code, use the Watch window, and use the Locals window.

Let us first introduce how to set breakpoints in VBA code. Setting a breakpoint allows us to halt a routine's execution, take a look at the value of any variables, and then continue execution. There several ways to set a breakpoint:

- Move the cursor to the statement where we want execution to stop and then press "F9."

- Move the cursor to the statement where we want execution to stop and then click the "Debug/Toggle Breakpoint."

- Click the gray margin to the left of the statement where we want execution to stop.

- Right-click the statement where we want execution to stop and click "Toggle/Breakpoint" in the pop-up menu.

When we execute a VBA procedure with a breakpoint, VBA goes into Break mode before the line with the breakpoint is executed. In Break mode, VBA stops running and the current statement is highlighted in bright yellow. In Break mode, we can do several things:

- Check values of variables in the Immediate window.

- Step through the code one line at a time by pressing "F8."

- See a variable's value by moving the mouse pointer over the variable.

- Skip the next statement or even go back a couple of statements.

- Edit a statement and then continue.

To illustrate how to debug VBA code in Break mode, we consider the following sub procedure:

```
1  Sub CopyTable()
2      Dim strSourceFile As String
3      Dim wbSource As Workbook
4      Dim wsDest As Worksheet
5      Dim i As Integer
6
7      strSourceFile = Application.GetOpenFilename()
8      If VarType(strSourceFile) <> vbString Then    '
           Cancel clicked
9          Exit Sub
10     End If
11
12     Set wbSource = Workbooks.Open(strSourceFile)
13     wbSource.Sheets("Sheet1").Range("A25:B134").
           Select
14
15     Set wsDest = ThisWorkbook.Sheets("Sheet1")
16     For i = 1 To Selection.Rows.Count
17         wsDest.Cells(i, 1).Value = Selection.Cells(i,
               1).Value
18         wsDest.Cells(i, 2).Value = Selection.Cells(i,
               2).Value
19     Next i
20
21     wbSource.Close
22 End Sub
```

The above sub procedure first gets a file name of a workbook from a dialog and then copies a range from the workbook to the active sheet of this workbook.

Suppose that we set a breakpoint at Line 19 of the code. If we run the code, VBA will enter Break model and the line with the breakpoint is highlighted, as shown in Figure 6.3. In Break mode, we can type the question mark "?" and a VBA statement to see the result. Figure 6.3 also shows the results of four VBA statements. From the figure we see that the value of the variable i is 1.

In some cases we want to see the values of variables near the end of a loop. If the loop has 100 iterations and we want to see the values of variables at the 90th iteration, then we need to respond to 90 prompts before the

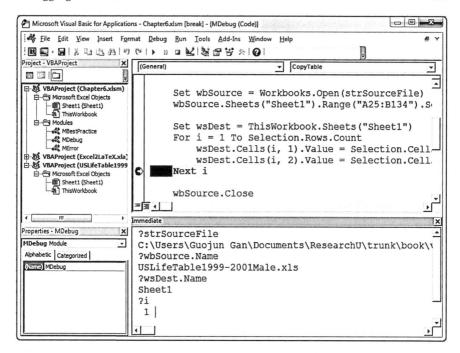

Figure 6.3: Checking values of VBA statements in the Immediate window in Break mode

code finally gets to the 90th iteration. A more efficient method to do this is to set a watch expression.

Suppose that we want to examine the code at the 20th iteration of the Fot-Next loop. We can set a watch expression that puts the procedure into Break mode. Figure 6.4 shows how to set such a watch expression. If we execute the code, VBA will enter into Break mode when the value of the variable i is equal to 20. In Break mode, we can check the values of variables in the Immediate window, as shown in Figure 6.5.

Another useful debugging tool is the Locals window, which can be used to look at the values of local variables in Break mode. To view the Locals window, we click the "View/Locals window." Figure 6.6 shows the Locals window in Break mode, which was triggered by the watch expression i=20. In the Locals window, we see the value and type of the variables.

Exercise 6.4. Consider the following sub procedure:

```
1  Sub GenerateRandNum()
```

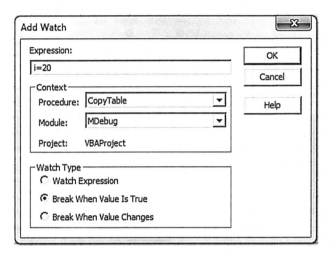

Figure 6.4: Setting a watch expression in the Watch window

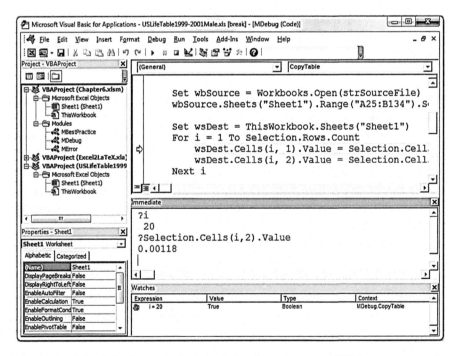

Figure 6.5: The Watches window and checking values of VBA statements in the Immediate window in Break mode

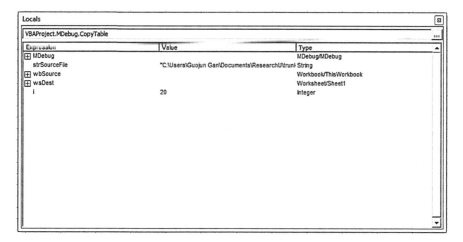

Figure 6.6: The Locals window in Break mode

```
2       Dim i As Integer
3       Dim dSum As Double
4       Dim dTemp As Double
5
6       Rnd (-1)
7       Randomize (0)
8       For i = 1 To 1000
9           dTemp = Rnd()
10          dSum = dSum + dTemp
11      Next i
12  End Sub
```

Do not modify the code. What is the value of the variable dTemp at the 900th iteration?

Debugging is an integrated part of the software development. It is difficult, if not impossible, to completely eliminate bugs in your VBA programs. Here are some tips that can help us keep bugs to a minimum (Walkenbach, 2013b):

- Use the Option Explicit statement at the beginning of modules.
- Format the VBA code with consistent indentation.
- Be careful with the On Error Resume Next statement.
- Use comments.

- Write simple Sub and Function procedures.
- Use the macro recorder to help identify properties and methods.
- Understand VBA's debugging tools.

6.3 Best Practices of VBA Coding

VBA best practices are useful for us to create high quality and reliable code. In this section, we present some best practices for Excel® VBA development. This section follows closely the third chapter of (Bovey et al., 2009) and presents best practices in the following three broad categories: naming convention, application structure, and application development.

Let us first present best practices for naming convention, which refers to the system we use to name various parts of our application. Table 6.3 gives some naming conventions for common elements of VBA programs.

Table 6.3: Naming Conventions

Element	Naming Convention
Variable	<scope><array><data type>[DescriptiveName]
Constant	<scope><data type>[DESCRIPTIVE_NAME]
User-defined type	Type [DESCRIPTIVE_NAME]
	<data type>[DescriptiveName]
	End Type
Enumeration type	Enum <project prefix>[GeneralDescr]
	<project prefix>[GeneralDescrSpecificName1]
	<project prefix>[GeneralDescrSpecificName2]
	End Enum
Module	M[GeneralDescr]
Class	C[GeneralDescr]
UserForm	F[GeneralDescr]

There are three scopes, and the scope specifiers are given in Table 6.4. If a variable is an array, the array specifier <array> is a. Otherwise, the array specifier is omitted (see Table 6.5). The specifiers of various data types are given in Table 6.6.

The following gives some examples of names based on the naming conventions:

Table 6.4: Scope Specifiers

Scope	Specifier
Public	g
Module-level	m
Procedure-level	nothing

Table 6.5: Array Specifiers

Array	Specifier
Array	a
Not an array	nothing

Table 6.6: Data Type Specifiers

Data Type	Specifier	Data Type	Specifier
Boolean	b	Worksheet	ws
Byte	byt	ComboBox	cbo
Currency	cur	CheckBox	chk
Date	dte	CommandButton	cmd
Decimal	dec	Frame	fra
Double	d	Label	lbl
Integer	i	ListBox	lst
Long	l	OptionButton	opt
Object	obj	SpinButton	spn
Single	sng	TextBox	txt
String	s	Variant	v
User-defined type	u	User-defined class variable	c
Chart	cht	UserForm variable	frm
Range	rng	VBA collection	col
Workbook	wb		

gsErrMsg A public `String` variable used to store an error message.

aiInd A procedure-level array of integers to hold an array of indices.

gdPI A public `Double` constant that holds the π value.

MDataAccess A module for data access.

We will see more examples in later chapters when we introduce case studies.

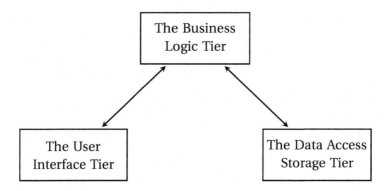

Figure 6.7: Three tiers of an application and their relationships

Next, let us introduce some best practices related to application structure and organization. Most Excel® applications have three logical tiers, shown in Figure 6.7. The user interface tier consists of all the code and visible elements your application uses to interact with users. The business logic tier is completely code-based and performs the core operations of your application. The data access and storage tier is responsible for storage and retrieval of data required by the application.

As we can see from Figure 6.7, the user interface tier and the data access and storage tier are not linked. The three tiers should by loosely coupled so that a significant change in one tier does not lead to significant changes to other two tiers.

VBA code should be organized into modules in a logical fashion. Procedures that perform similar functions should be grouped into the same code module. Each procedure should be responsible for a well-defined task that is easy to understand, validate, document, and maintain. Important guidelines for creating procedures include:

Encapsulation A procedure should completely encapsulate the logical operation it performs.

No duplicate code If a piece of code is used in multiple places to perform the same operation, then you should extract the piece of code and place it into a separate procedure.

Procedure size reduction If a procedure contains more than 100 lines of code, it should be factored into multiple single-purpose procedures.

Limiting the number of arguments A procedure should accept no more than five arguments.

Finally let us introduce some best practice for general application development in the following areas: code commenting, code readability, use of module directives, use of variables and constants, defensive coding, version control, and change documentation.

Good code commenting is an important practice in programming. In VBA programming, there are three categories of comments: module-level comments, procedure-level comments, and internal comments. A good module-level comment should be placed at the top of the module and look something like the following example:

```
1  '
2  ' Description: A brief description of the purpose of
3  '              the code in this module.
4  '
5  ' Revision history:
6  ' Date          Developer          Action
7  '
   ' -------------------------------------------------------------
8  ' 01/30/2016   Guojun Gan         Created
9  '
10 Option Explicit
```

Procedure-level comments are usually the most detailed comments in your application. The comment of a procedure is placed at the top of the procedure and contains the following content: the purpose of the procedure, usage, a detailed list of arguments, a description of expected return values if the procedure is a function procedure, and changes made to the procedure. Internal comments are used within the body of the code itself to describe the purpose of any code that is not self-evident. It is important to make sure that comments are updated when the underlying code is changed.

Writing readable code is another important practice in programming. Although code layout does not make any difference to computers, good visual layout of code allows us to infer lots of information about the logic structure of the program. Related code should be grouped together, and unrelated code should be separated by blank lines. Indentation should be used to indicate a related block of code.

In addition to good code commenting and writing readable code, the following is a list of best practices in other areas, such as use of directives and use of variables and constants:

Option Explicit Alway use the Option Explicit in every module. This statement forces you to explicitly declare all the variables you use.

Option Private Module The Option Private Module statement makes all procedures within the module unavailable to other modules or the Excel user interface. This statement should be used to hide procedures that should not be called from outside of your application.

Option Base 1 The Option Base 1 statement causes the lower bound of all array variables whose lower bound has not been specified to have a lower bound of 1. This statement should not be used.

Option Compare Text The Option Compare Text statements causes all string comparisons within the module to be text-based rather than binary. This statement should be avoided.

Reuse variables Avoid reusing variables. Each variable should serve one purpose only.

Variant data type Avoid using the Variant data type for the following reasons: first, variants are not efficient; second, data stored in a variant can behave unexpectedly.

Type conversion VBA automatically converts one data type to another data type. Do not mix variables of different data types in your VBA expressions without using the explicit casting functions (e.g., CStr and CDbl).

Array bounds Never hand-code the array bounds in a loop. Always use the LBound function and the UBound function to get the bounds of an array.

ByRef and ByVal There are two conventions to declare arguments of a procedure: ByRef and ByVal. If an argument is declared with the ByRef convention, then the memory address of the variable rather than the value of the variable is passed. In such cases, modification of the variable will be visible in the calling procedure. If an argument is declared with the `ByVal` convention, the value of the variable is passed. Always explicitly declare procedure arguments as `ByRef` or `ByVal`.

Argument validation Always validate arguments before using them in procedures so that erroneous input is caught as soon as possible.

Infinite loops Always add a guard counter to terminate an infinite loop when the number of loops executed exceeds the highest number that should ever occur in practice.

Use CodeNames to reference sheet objects Always reference worksheets and chart sheets by their CodeNames. It is risky to identify sheets by their tab names, because the tab names can be changed during execution of the program.

Validate the data types of selection Always check the object type of the selection using the `TypeName` function or the `TypeOf-Is` structure.

Saving versions Use a version control system that allows you to recover an earlier version of your project if you have encountered problems with your current version.

For detailed discussion of the above best practices, readers are referred to (Bovey et al., 2009, Chapter 3).

6.4 Summary

In this chapter, we introduced how to handle runtime errors and how to use VBA's debugging tools to identify potential logic errors. We also introduced some best practices of VBA coding. For more information about best practices, readers are referred to (Bovey et al., 2009).

Part II

Applications

7

Generating Payment Schedules

In this chapter, we implement a simple Visual Basic for Applications (VBA) program that can generate regular payment dates of an interest rate swap according to a business calendar. Figure 7.1 shows the interface of the payment schedule generator.

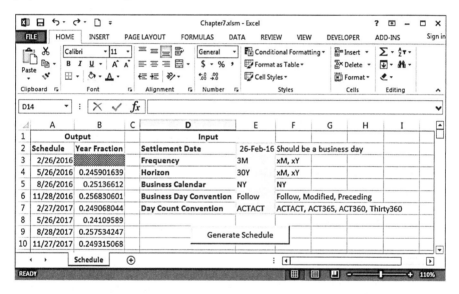

Figure 7.1: Interface of the payment schedule generator

From 7.1, we see that the interface contains six input cells highlighted in yellow and one button. The output is shown in the first two columns. The first column shows the payment dates, and the second column shows the year fractions between these dates. The goal of this chapter is to implement the payment generator.

7.1   Introduction

A derivative (Garrett, 2013) is a concept in finance and refers to a contract that derives its value from the performance of an underlying entity, such as an asset, index, or interest rate. Derivatives have been used for risk management purposes. For example, insurance companies have used interest rate swaps to hedge the interest rate risks associated with variable annuity contracts.

An interest rate swap (IRS) is a financial derivative in which two parties agree to exchange interest rate cash flows, based on a specified notional amount. Under an IRS, each counterpart agrees to pay either a fixed or floating rate denominated in a particular currency to the other counterpart.

When pricing an IRS, one of the first steps is to generate a schedule of dates, based on the holiday calendar and the date rolling rules specified in the contract. The schedule of dates is used to project cash flows of the IRS. Projecting cash flows is also dependent on the day count convention specified in the swap contract.

In this chapter, we introduce how to generate a schedule of dates based on the US holiday calendar. We will also introduce some day count conventions and date rolling rules.

7.2   Public Holidays in the United States

There are ten public holidays and observances in the United States. The New York Stock Exchange (NYSE) markets observe these holidays. Theses holidays are listed in Table 7.1.

The holidays can be divided into two groups in terms of how they are determined. The first group includes New Year's Day, Independence Day, Veterans Day, and Christmas Day. The holidays in the first group occur on a specified date. If a specified date falls on a weekend, then either the previous Friday or the next Monday is observed. The second group includes the remaining holidays. The holidays in the second group are determined based on a particular weekday.

Table 7.1: Holidays and Observances in the United States

Name	Date
New Year's Day	January 1
	(January 2 if January 1 is a Sunday,
	December 31 if January 1 is a Saturday)
Martin Luther King Jr. Day	The third Monday of January
Washington's Birthday	The third Monday of February
Memorial Day	The last Monday of May
Independence Day	July 4
	(July 5 if July 4 is a Sunday,
	July 3 if July 4 is a Saturday)
Labor Day	The first Monday of September
Columbus Day	The second Monday of October
Veterans Day	November 11
	(November 12 if November 11 is a Sunday,
	November 10 if November 11 is a Saturday)
Thanksgiving Day	The fourth Thursday of November
Christmas Day	December 25
	(December 26 if December 25 is a Sunday,
	December 24 if December 25 is a Saturday)

7.3   Julian and Gregorian Calendars

A calender is a system for assigning a unique date to every day so that the order of the dates is obvious and the number of days between two dates can be calculated easily (Richards, 2012). There are about forty calendars in use in the world today, and most calendars are based on astronomical cycles. In this section, we introduce the Julian and Gregorian calendars and algorithms for converting a date between the two calendars.

The Julian calendar was introduced by Julius Caesar in 46 BC. The Julian calendar was the predominant calendar in the Roman world and most of Europe until it was replaced by the Gregorian calendar. The average length of the year in the Julian calendar is 365.25 days.

The Gregorian calendar is a solar calendar, which maintains accurate synchrony with the astronomical year but does not keep the months in synchrony with the lunation (the cycle of the phases of the moon) of the moon. The Gregorian calendar is the official calendar of the United Kingdom since

1752 and is used throughout the world for secular purpose. The average length of the year in the Gregorian calendar is 365.2425 days.

A date consists of three components: a year, a month, and a day of the month. The three components correspond to the three principle astronomical cycles: the revolution of the Earth around the sun (a year), the revolution of the moon around the Earth (a month), and the rotation of the Earth on its axis (a day). The complexity of calendars comes from the fact that a month and a year do not comprise an integral number of days, and they are neither constant nor perfectly commensurable with each other.

Table 7.2: Parameters for Calendar Conversion Algorithms

Parameter	Value	Parameter	Value	Parameter	Value
y	4716	p	1461	t	2
j	1401	q	0	w	2
m	2	v	3	A	184
n	12	u	5	B	274277
r	4	s	153	C	−38

There are algorithms for interconverting dates in the Gregorian calendar and the Julian day number (Richards, 2012). Let **D/M/Y** be a date in the Gregorian calendar. Then, the corresponding Julian day number **J** is calculated as follows:

(a) $h = \mathbf{M} - m$

(b) $g = \mathbf{Y} + y - (n - h)\backslash n$

(c) $f = \text{mod}\,(h - 1 + n, n)$

(d) $e = (pg + q)\backslash r + \mathbf{D} - 1 - j$

(e) $\mathbf{J} = e + (sf + t)\backslash u - (3 * ((g + A)\backslash 100))\backslash 4 - C,$

where, mod (A, B) represents the remainder when A is divided by B, and \ denotes integer division in which any remainder is ignored.

To convert a Julian day number **J** to a date **D/M/Y** in the Gregorian calendar, we use the following algorithm:

(a) $f = \mathbf{J} + j + (((4\mathbf{J} + B)\backslash 146097) \times 3)\backslash 4 + C$

(b) $e = rf + v$

(c) $g = \text{mod}\,(e, p)\backslash r$

(d) $h = ug + w$

(e) $\mathbf{D} = \mathrm{mod}\,(h, s)\backslash u + 1$

(f) $\mathbf{M} = \mathrm{mod}\,(h\backslash s + m, n) + 1$

(g) $\mathbf{Y} = e\backslash p - y + (n + m - \mathbf{M})\backslash n$

The parameters used in the above two algorithms are given in Table 7.2.

7.4 Day Count Conventions

A day count convention determines how interest accrues over time for interest-earning investments. A day count convention is also used to quantify periods of time when discounting a cash flow to its present value.

In this section, we introduce the definition of the following day count conventions: ACT/ACT, ACT/365, ACT/360, and 30/360. Since there is not a unique definition of day count conventions, we follow the International Swaps and Derivatives Association (ISDA), which has documented some day count conventions (ISDA, 2006).

The ACT/ACT day count convention is defined as follows:

$$\Delta = \frac{\text{Days not in leap year}}{365} + \frac{\text{Days in leap year}}{366}, \tag{7.1}$$

where the days in the numerators are calculated on a Julian day difference basis, and a leap year is a year that contains 366 days by extending February to 29 days. In the Gregorian calendar, a year $\mathbf{Y}$ is a leap year if $\mathbf{Y}$ satisfied one of the following two conditions:

(a) $\mathbf{Y}$ is divisible by 4 and is not divisible by 100

(b) $\mathbf{Y}$ is divisible by 400

Exercise 7.1. Which of the following years are leap years?

$$1900, 1996, 2000, 2016, 2100.$$

The ACT/365 day count convention is defined as follows:

$$\Delta = \frac{\text{Days in the calculation period}}{365}, \tag{7.2}$$

where the days in the numerator are calculated on a Julian day difference basis. Similarly, the ACT/360 day count convention is defined as follows:

$$\Delta = \frac{\text{Days in the calculation period}}{360}, \tag{7.3}$$

where the days in the numerator are calculated on a Julian day difference basis.

The 30/360 day count convention is defined as follows:

$$\Delta = \frac{360(Y_2 - Y_1) + 30(M_2 - M_1) + (D_2 - D_1)}{360}, \tag{7.4}$$

where

- Y_1 is the year in which the first day of the calculation period falls;
- Y_2 is the year in which the day immediately following the last day included in the calculation period falls;
- M_1 is the month in which the first day of the calculation period falls;
- M_2 is the month in which the day immediately following the last day included in the calculation period falls;
- D_1 is the first day of the calculation period. If the first day is 31, then $D_1 = 30$;
- D_2 is the day immediately following the last day included in the calculation period. If $D_1 = 30$ and the day immediately following the last day is 31, then $D_2 = 30$.

7.5   Business Day Conventions

When a payment date falls on a holiday or weekend according to a specified business calendar, we need to adjust the payment date forward or backward in time, such that it falls on a business day in the same business calendar. A business day convention refers to a convention for adjusting any payment date that falls on a nonbusiness day (e.g., holidays or weekends). There are three business day conventions (ISDA, 2006): following, modified following, and preceding.

If "following" is specified, the payment date is adjusted to the next business day. If "modified following" is specified, the payment date is adjusted to the next business day unless next business day is in the next calendar month, in which case the payment date is adjusted to the previous business day. If "preceding" is specified, the payment date is adjusted to the previous business day.

Exercise 7.2. The 2016 holidays observed by the NYSE markets are: January 1, January 18, February 15, March 25, May 30, July 4, September 5, November 24, and December 26. Adjust the following payment dates according to various business day conventions:

Payment Date	Following	Modified Following	Preceding
January 16, 2016			
May 30, 2016			
November 25, 2016			

7.6 Implementation

In this section, we implement the payment schedule generator with the interface shown in Figure 7.1. We would like to put the VBA code into three modules: MInterface, MDate, and MHoliday. The module MInterface contains VBA code for the button. The module MDate contains VBA code for manipulating dates. The module MHoliday contains code for handling business calendars and generating payment date schedules.

7.6.1 MDate Module

Let us first implement the module for manipulating dates. We name this module as MDate. At the declarations section of this module, we define some constants given in Table 7.2 as follows:

```
1  Option Explicit
2
3  Const cony As Long = 4716
4  Const conj As Long = 1401
5  Const conm As Long = 2
```

```
 6 Const conn As Long = 12
 7 Const conr As Long = 4
 8 Const conp As Long = 1461
 9 Const conq As Long = 0
10 Const conv As Long = 3
11 Const conu As Long = 5
12 Const cons As Long = 153
13 Const cont As Long = 2
14 Const conw As Long = 2
15 Const conA As Long = 184
16 Const conB As Long = 274277
17 Const conC As Long = -38
```

In the above code, we named the constants with the prefix "con" plus the parameter names given in Table 7.2 so that we can easily connect the VBA constants to the parameters. These constants are used to interconvert Gregorian dates and Julian day numbers.

Once we have defined the constants in the module, we can implement the function for converting a Gregorian date to a Julian day number. Following the algorithm given in Section 7.3, we implement the function as follows:

```
 1 Function ToJulian(ByVal datIn As Date) As Long
 2     Dim lD As Long
 3     Dim lM As Long
 4     Dim lY As Long
 5
 6     Dim lh As Long
 7     Dim lg As Long
 8     Dim lf As Long
 9     Dim le As Long
10
11     lD = Day(datIn)
12     lM = Month(datIn)
13     lY = Year(datIn)
14
15     lh = lM - conm
16     lg = lY + cony - (conn - lh) \ conn
17     lf = (lh - 1 + conn) Mod conn
18     le = (conp * lg + conq) \ conr + lD - 1 - conj
19     ToJulian = le + (cons * lf + cont) \ conu - (3 *
           ((lg + conA) \ 100)) \ 4 - conC
20 End Function
```

In the above code, we used the integer division operator \.

Similarly, we can follow the algorithm to implement the function for converting a Julian day number to a Gregorian date as follows:

```
1  Function FromJulian(ByVal lJ As Long) As Date
2      Dim lD As Long
3      Dim lM As Long
4      Dim lY As Long
5
6      Dim lh As Long
7      Dim lg As Long
8      Dim lf As Long
9      Dim le As Long
10
11     lf = lJ + conj + (((4 * lJ + conB) \ 146097) * 3) _
           \ 4 + conC
12     le = conr * lf + conv
13     lg = (le Mod conp) \ conr
14     lh = conu * lg + conw
15     lD = (lh Mod cons) \ conu + 1
16     lM = ((lh \ cons + conm) Mod conn) + 1
17     lY = le \ conp - cony + (conn + conm - lM) \ conn
18     FromJulian = DateSerial(lY, lM, lD)
19 End Function
```

Before we continue to implement other functions, let us test the functions ToJulian and FromJulian. We can write a sub procedure to test these two functions as follows:

```
1  Sub TestJulianGregorian()
2      Dim datA As Date
3      Dim lJ As Long
4
5      datA = #1/1/1998#
6      lJ = ToJulian(datA)
7
8      Debug.Print lJ
9      Debug.Print FromJulian(lJ)
10
11     datA = #5/1/2016#
12     lJ = ToJulian(datA)
13
14     Debug.Print lJ
15     Debug.Print FromJulian(lJ)
16 End Sub
```

In the above sub procedure, we first converted a date to a Julian day number and then converted the Julian day number back to a date. Executing the above sub procedure produces the following output in the Immediate window:

```
1  2450815
2 1/1/1998
3  2457510
4 5/1/2016
```

From the above output, we see that we can recover the dates exactly from their Julian day numbers.

Now, let us implement functions for adding intervals (e.g., days, months, years) to a date. Let us first implement the function AddDays, which returns a date by adding a given number of days to an input date. We can implement the function as follows:

```
1 Function AddDays(ByVal datIn As Date, ByVal lCount As
     Long) As Date
2     Dim lJ As Long
3
4     lJ = ToJulian(datIn) + lCount
5     AddDays = FromJulian(lJ)
6 End Function
```

To add days to a date, we first convert the date to a Julian day number and add the given number of days to the Julian day number. Then we convert the resulting Julian day number to a Gregorian date.

To test the function AddDays, we use the following sub procedure:

```
1 Sub TestAddDays()
2     Debug.Print AddDays(#2/1/2016#, 28)
3     Debug.Print AddDays(#2/1/2016#, 365)
4     Debug.Print AddDays(#2/1/2016#, -365)
5 End Sub
```

Executing the above code gives the following output:

```
1 2/29/2016
2 1/31/2017
3 2/1/2015
```

The above output shows that the function AddDays works as expected.

Adding months or years to a date is more complicated than adding days, in that the number of days in a month is not the same. For example, adding 12 months or 1 year to 2/29/2016 is not 2/29/2017 because the year 2017 is not a leap year. We need a function to get the maximum number of days in a month. We can implement the function as follows:

```
1 Function DaysInMonth(ByVal datIn As Date) As Integer
2     Dim lY As Long
3     Dim lM As Long
4
5     lY = Year(datIn)
6     lM = Month(datIn)
7     DaysInMonth = DateSerial(lY, lM + 1, 1) -
          DateSerial(lY, lM, 1)
8 End Function
```

The function DaysInMonth returns the number of days in the month of the input date. The number of days is calculated by subtracting two dates created by the function DateSerial, which is a VBA built-in function. We can test the function as follows:

```
1 Sub TestDaysInMonth()
2     Dim i As Integer
3     For i = 1 To 12
4         Debug.Print DaysInMonth(DateSerial(2016, i,
              1));
5     Next i
6     Debug.Print ""
7     For i = 1 To 12
8         Debug.Print DaysInMonth(DateSerial(2017, i,
              1));
9     Next i
10 End Sub
```

In the above test function, we used the Fot-Next loop to display the number of days in each month of years 2016 and 2017. Executing the above code gives:

```
1  31   29   31   30   31   30   31   31   30   31   30   31
2  31   28   31   30   31   30   31   31   30   31   30   31
```

From the output, we see that the function calculates the number of days correctly.

Then we can implement the function for adding months as follows:

```
 1 Function AddMonths(ByVal datIn As Date, ByVal lCount
       As Long) As Date
 2     Dim lD As Long
 3     Dim lM As Long
 4     Dim lY As Long
 5     Dim lDMax As Integer
 6     Dim datTmp As Date
 7
 8     lD = Day(datIn)
 9     lM = Month(datIn)
10     lY = Year(datIn)
11
12     lM = lM + lCount
13     Do While lM > 12 ' addition of months
14          lY = lY + 1
15          lM = lM - 12
16     Loop
17
18     Do While lM < 1 ' subtraction of months
19          lY = lY - 1
20          lM = lM + 12
21     Loop
22
23     ' need to truncate days in month
24     datTmp = DateSerial(lY, lM, 1)
25     lDMax = DaysInMonth(datTmp)
26     If lD > lDMax Then
27          lD = lDMax
28     End If
29     AddMonths = DateSerial(lY, lM, lD)
30 End Function
```

The function AddMonths returns a date by adding a given number of months to an input date. We truncate some days, if necessary. To test this function, we use the following sub procedure:

```
 1 Sub TestAddMonths()
 2     Debug.Print AddMonths(DateSerial(2015, 11, 30),
           3)
 3     Debug.Print AddMonths(DateSerial(2015, 2, 28),
           12)
 4     Debug.Print AddMonths(DateSerial(2015, 3, 30),
           -5)
 5 End Sub
```

Executing the above code gives

```
1 2/29/2016
2 2/28/2016
3 10/30/2014
```

The above output shows that the function AddMonths works correctly.

Similarly, we can implement the function for adding years as follows:

```
1 Function AddYears(ByVal datIn As Date, ByVal lCount
      As Long) As Date
2     Dim lD As Long
3     Dim lM As Long
4     Dim lY As Long
5     Dim lDMax As Integer
6     Dim datTmp As Date
7
8     lD = Day(datIn)
9     lM = Month(datIn)
10    lY = Year(datIn) + lCount
11
12    ' handle Feb 29
13    datTmp = DateSerial(lY, lM, 1)
14    lDMax = DaysInMonth(datTmp)
15    If lD > lDMax Then
16        lD = lDMax
17    End If
18    AddYears = DateSerial(lY, lM, lD)
19 End Function
```

The function AddYears returns a date by adding a given number of years to an input date. It is similar to the function AddMonths. But this function is simpler than the function AddMonths, as we only need to take care of leap years. We can test this function as follows:

```
1 Sub TestAddYears()
2     Debug.Print AddYears(DateSerial(1998, 1, 1), 18)
3     Debug.Print AddYears(DateSerial(2000, 2, 29), 15)
4     Debug.Print AddYears(DateSerial(2016, 2, 29), 1)
5 End Sub
```

Executing the above sub procedure gives

```
1 1/1/2016
2 2/28/2015
3 2/28/2017
```

From the output, we see that the function AddYears works well.

We can combine the functions for adding intervals into a single function as follows:

```
 1 Function OffsetDate(ByVal datIn As Date, ByVal lCount
      As Long, ByVal sPeriod As String) As Date
 2     Select Case sPeriod
 3         Case "M"
 4             OffsetDate = AddMonths(datIn, lCount)
 5         Case "Y"
 6             OffsetDate = AddYears(datIn, lCount)
 7         Case Else
 8             Err.Raise Number:=1003, Source:="MHoliday
                  .GenerateDate", Description:="Invalid
                  period string"
 9     End Select
10 End Function
```

The function OffsetDate returns a date by adding a given number of periods to an input date. In the current version, we can only add months or years to a date. If the input period is not "M" or "Y," the function will raise an error.

Exercise 7.3. The function OffsetDate can handle only two frequencies: "M" and "Y." Modify the function OffsetDate so that it can handle the following additional frequencies: "D," "W," and "Q," where D, W, and Q refer to day, week, and quarter.

Now, let us implement functions for calculating the year fractions based on the four day count conventions. We first implement a function that tells whether a year is a leap year as follows:

```
 1 Function IsLeapYear(ByVal lY As Long) As Boolean
 2     Dim bRes As Boolean
 3
 4     bRes = False
 5     If lY Mod 4 = 0 And lY Mod 100 <> 0 Then
 6         bRes = True
```

```
7      End If
8      If lY Mod 400 = 0 Then
9          bRes = True
10     End If
11
12     IsLeapYear = bRes
13 End Function
```

The function `IsLeapYear` returns true if an input year is a leap year in the Gregorian calendar. We can test this function using the following sub procedure:

```
1 Sub TestIsLeapYear()
2     Dim i As Integer
3
4     For i = 2000 To 2009
5         Debug.Print IsLeapYear(i) & " ";
6     Next i
7 End Sub
```

Executing the above sub procedure gives

```
1 True False False False True False False False True
      False
```

The output shows that the function `IsLeapYear` works as expected.

Then, we implement a function to calculate the number of days in leap years between two given dates. We can implement this function as follows:

```
1 Function DaysInLeapYears(ByVal datA As Date, ByVal
      datB As Date) As Long
2     ' If datA is before datB, we switch the order
3     If ToJulian(datA) > ToJulian(datB) Then
4         DaysInLeapYears = DaysInLeapYears(datB, datA)
5         Exit Function
6     End If
7
8     Dim l As Long
9     Dim lYA As Long
10    Dim lYB As Long
11    Dim lDays As Long
12    ' Then we can assume datA is before datB
13    lYA = Year(datA)
14    lYB = Year(datB)
15
```

```
16      For l = lYA To lYB
17          If IsLeapYear(l) Then
18              lDays = lDays + _
19                  ToJulian(DateSerial(l + 1, 1, 1)) - _
20                  ToJulian(DateSerial(l, 1, 1))
21          End If
22      Next l
23      If IsLeapYear(lYA) Then
24          lDays = lDays + ToJulian(DateSerial(lYA - 1,
                12, 31)) - ToJulian(datA)
25      End If
26
27      If IsLeapYear(lYB) Then
28          lDays = lDays + ToJulian(datB) - ToJulian(
                DateSerial(lYB, 12, 31))
29      End If
30
31      DaysInLeapYears = lDays
32 End Function
```

The idea behind the above implementation is that we add all the days in leap years between the years of the two given dates. Then, we subtract the days outside the two given dates. To this this function, we use the following sub procedure:

```
1 Sub TestDaysInLeapYears()
2      Debug.Print DaysInLeapYears(#1/1/2016#,
           #1/31/2016#)
3      Debug.Print DaysInLeapYears(#1/1/2016#,
           #1/31/2018#)
4      Debug.Print DaysInLeapYears(#1/1/2015#,
           #1/31/2018#)
5      Debug.Print DaysInLeapYears(#1/1/2015#,
           #12/31/2015#)
6 End Sub
```

Executing the above sub procedure gives

```
1   30   365   366   0
```

The output shows that the function calculates the number of days in leap years correctly.

Using the function DaysInLeapYears, we can calculate the year fraction based on the ACT/ACT convention as follows:

```
 1 Function ActAct(ByVal datA As Date, ByVal datB As
     Date) As Double
 2     Dim dDaysNonLeap As Double
 3     Dim dDaysLeap As Double
 4     Dim lYA As Long
 5     Dim lYB As Long
 6     Dim l As Long
 7
 8     dDaysLeap = DaysInLeapYears(datA, datB)
 9     dDaysNonLeap = Abs(ToJulian(datA) - ToJulian(datB
         )) - dDaysLeap
10
11     ActAct = dDaysLeap / 366 + dDaysNonLeap / 365
12 End Function
```

In the above function, the number of days in nonleap years is calculated as the total number of days minus the number of days in leap years.

To calculate the year fraction between two dates based on the 30/360 day count convention, we implement the following function:

```
 1 Function Thirty360(ByVal datA As Date, ByVal datB As
     Date) As Double
 2     Dim lY1 As Long
 3     Dim lM1 As Long
 4     Dim lD1 As Long
 5     Dim lY2 As Long
 6     Dim lM2 As Long
 7     Dim lD2 As Long
 8
 9     If ToJulian(datA) > ToJulian(datB) Then
10         Thirty360 = Thirty360(datB, datA)
11         Exit Function
12     End If
13
14     lY1 = Year(datA)
15     lM1 = Month(datA)
16     lD1 = Day(datA)
17     lY2 = Year(datB)
18     lM2 = Month(datB)
19     lD2 = Day(datB)
20     If lD1 = 31 Then
21         lD1 = 30
22     End If
23     If lD1 = 30 And lD2 = 31 Then
```

```
24        1D2 = 30
25     End If
26     Thirty360 = 1Y2 - 1Y1 + (1M2 - 1M1) / 12# + (1D2
          - 1D1) / 360#
27 End Function
```

The sign # after an integer tells VBA to treat the integer as a double.

We can combine the above two functions in the following function:

```
1 Function TFrac(ByVal datA As Date, ByVal datB As Date
     , ByVal sDCC As String) As Double
2     Dim 1JA As Long
3     Dim 1JB As Long
4     1JA = ToJulian(datA)
5     1JB = ToJulian(datB)
6
7     If 1JA > 1JB Then
8         TFrac = TFrac(datB, datA, sDCC)
9         Exit Function
10    End If
11
12    Select Case sDCC
13        Case "ACTACT"
14            TFrac = ActAct(datA, datB)
15        Case "ACT365"
16            TFrac = (1JB - 1JA) / 365#
17        Case "ACT360"
18            TFrac = (1JB - 1JA) / 360#
19        Case "Thirty360"
20            TFrac = Thirty360(datA, datB)
21        Case Else
22            Err.Raise Number:=1007, Source:="MDate.
                 TFrac", Description:="Unknown day
                 count convention."
23    End Select
24 End Function
```

The function TFrac calculates the year fraction between two dates according to a specified day count convention. The function will raise an error if an unknown day count convention is encountered. We can test this function as follows:

```
1 Sub TestTFrac()
2     Dim datA As Date
3     Dim datB As Date
```

```
 4
 5     datA = #12/31/2015#
 6     datB = #12/31/2016#
 7
 8     Debug.Print TFrac(datA, datB, "ACTACT")
 9     Debug.Print TFrac(datA, datB, "ACT365")
10     Debug.Print TFrac(datA, datB, "ACT360")
11     Debug.Print TFrac(datA, datB, "Thirty360")
12 End Sub
```

Executing the above test function gives the following output:

```
1  1
2  1.0027397260274
3  1.01666666666667
4  1
```

The output shows that the function works as expected.

7.6.2 MHoliday Module

In this subsection, we implement a module with functions for generating payment schedules based on the US business calendar. We name this module MHoliday.

The first function we would like to implement is one that checks whether a date is a weekend. We implement this function as follows:

```
1 Function IsWeekend(ByVal datIn As Date) As Boolean
2     Dim iWd As Integer
3     iWd = Weekday(datIn)
4     If iWd = vbSaturday Or iWd = vbSunday Then
5         IsWeekend = True
6     Else
7         IsWeekend = False
8     End If
9 End Function
```

The function IsWeekend returns true if the input date is a weekend. We can test this function as follows:

```
1 Sub TestIsWeekend()
2     Dim i As Integer
3     For i = 1 To 7
4         Debug.Print IsWeekend(DateSerial(2016, 5, i))
             & " ";
```

```
5     Next i
6 End Sub
```

Executing the above sub procedure gives

```
1 True False False False False False True
```

The output shows that the function works as expected.

The second function checks whether a date is a US holiday. We implement this function as follows:

```
1 Function IsHolidayNY(ByVal datIn As Date) As Boolean
2     If IsWeekend(datIn) Then
3         IsHolidayNY = False
4         Exit Function
5     End If
6
7     Dim iWd As Integer
8     Dim iM As Integer
9     Dim iDoM As Integer
10    Dim bRet As Boolean
11
12    bRet = False
13    iWd = Weekday(datIn)
14    iM = Month(datIn)
15    iDoM = Day(datIn)
16
17    ' January 1
18    If (iM = 1 And iDoM = 1) Or _
19        (iM = 12 And iDoM = 31 And iWd = vbFriday) Or
        _
20        (iM = 1 And iDoM = 2 And iWd = vbMonday) Then
21        bRet = True
22    End If
23    ' Martin Luther King, Third Monday of January
24    If iM = 1 And iWd = vbMonday And iDoM > 14 And
          iDoM < 22 Then
25        bRet = True
26    End If
27    ' Washington's Birthday,  The third Monday of
          February
28    If iM = 2 And iWd = vbMonday And iDoM > 14 And
          iDoM < 22 Then
29        bRet = True
30    End If
```

```
31     ' Memorial Day, The last Monday of May
32     If iM = 5 And iWd = vbMonday And iDoM > 24 Then
33        bRet = True
34     End If
35     ' Independence Day, July 4
36     If (iM = 7 And iDoM = 4) Or _
37        (iM = 7 And iDoM = 3 And iWd = vbFriday) Or _
38        (iM = 7 And iDoM = 5 And iWd = vbMonday) Then
39        bRet = True
40     End If
41     ' Labor Day, The first Monday of September
42     If iM = 9 And iWd = vbMonday And iDoM < 8 Then
43        bRet = True
44     End If
45     ' Columbus Day, The second Monday of October
46     If iM = 10 And iWd = vbMonday And iDoM > 7 And
          iDoM < 15 Then
47        bRet = True
48     End If
49     ' Veterans Day, November 11
50     If (iM = 11 And iDoM = 11) Or _
51        (iM = 11 And iDoM = 10 And iWd = vbFriday) Or
           _
52        (iM = 11 And iDoM = 12 And iWd = vbMonday)
              Then
53        bRet = True
54     End If
55     ' Thanksgiving Day, The fourth Thursday of
          November
56     If iM = 11 And iWd = vbThursday And iDoM > 21 And
          iDoM < 29 Then
57        bRet = True
58     End If
59     ' Christmas Day, December 25
60     If (iM = 12 And iDoM = 25) Or _
61        (iM = 12 And iDoM = 24 And iWd = vbFriday) Or
           _
62        (iM = 12 And iDoM = 26 And iWd = vbMonday)
              Then
63        bRet = True
64     End If
65
66     IsHolidayNY = bRet
67 End Function
```

The function `IsHolidayNY` returns true if the input date is a public holiday in the United States. Since a weekend cannot be a holiday, the function returns false if the input date falls on a weekend. This function is implemented according to the public holidays given in Table 7.1. We can test this function as follows:

```
1  Sub TestIsHolidayNY()
2      Dim datA As Date
3      Dim i As Integer
4
5      For i = 0 To 365
6          datA = AddDays(DateSerial(2016, 1, 1), i)
7          If IsHolidayNY(datA) Then
8              Debug.Print datA
9          End If
10     Next i
11
12 End Sub
```

The above sub procedure displays all the holidays in 2016. Executing the sub procedure gives the following output:

```
1  1/1/2016
2  1/18/2016
3  2/15/2016
4  5/30/2016
5  7/4/2016
6  9/5/2016
7  10/10/2016
8  11/11/2016
9  11/24/2016
10 12/26/2016
```

If we check the US holidays in the Internet, we see that the above list of holidays is correct.

The third function in this module checks whether a date is a business day based on a calendar. We implement this function as follows:

```
1  Function IsBusinessDay(ByVal datIn As Date, ByVal
       sCalendar As String) As Boolean
2      Dim bRet As Boolean
3      bRet = False
4
5      Select Case sCalendar
6          Case "NY"
```

```
 7              bRet = (Not IsWeekend(datIn)) And (Not
                   IsHolidayNY(datIn))
 8          Case Else
 9              Err.Raise Number:=1001, Source:="MHolidy.
                   IsBusinessDay", Description:="Unknown
                   calendar."
10      End Select
11
12      IsBusinessDay = bRet
13 End Function
```

The function IsBusinessDay returns true if the input date is a business day according to a given business calendar. In the current version, only one business calendar (i.e., NY) is implemented. If the input business calendar is not "NY," the procedure will raise an error by calling the VBA function Err.Raise. To test this function, we use the following sub procedure:

```
1 Sub TestIsBusinessDay()
2     Debug.Print IsBusinessDay(#2/15/2016#, "NY")
3     Debug.Print IsBusinessDay(#12/26/2016#, "NY")
4     Debug.Print IsBusinessDay(#12/27/2016#, "NY")
5 End Sub
```

Executing the above test procedure gives

```
1 False
2 False
3 True
```

The output shows that the function works as expected.

7.6.3 MSchedule Module

The module MSchedule contains functions for generating payment schedules according to a specified business calendar, a specified payment frequency, and a specified payment horizon.

Since a payment date should be a business day, we need a function for adjusting a payment date based on a given business day convention. We can implement this function as follows:

```
1 Function RollDate(ByVal datIn As Date, ByVal sBDC As
     String, ByVal sCalendar As String) As Date
2     Dim datTmp As Date
3
```

```
 4      datTmp = datIn
 5      Select Case sBDC
 6          Case "Follow"
 7              Do While Not IsBusinessDay(datTmp,
                    sCalendar)
 8                  datTmp = AddDays(datTmp, 1)
 9              Loop
10          Case "Modified"
11              Do While Not IsBusinessDay(datTmp,
                    sCalendar)
12                  datTmp = AddDays(datTmp, 1)
13              Loop
14              If Month(datTmp) <> Month(datIn) Then
15                  datTmp = datIn
16                  Do While Not IsBusinessDay(datTmp,
                        sCalendar)
17                      datTmp = AddDays(datTmp, -1)
18                  Loop
19              End If
20          Case "Preceding"
21              Do While Not IsBusinessDay(datTmp,
                    sCalendar)
22                  datTmp = AddDays(datTmp, -1)
23              Loop
24          Case Else
25              Err.Raise Number:=1006, Source:="MHoliday
                    .RollDate", Description:="Unknown
                    business day convention."
26      End Select
27
28      RollDate = datTmp
29 End Function
```

The function RollDate adjusts a payment date that falls on a nonbusiness day (i.e., a weekend or holiday) according to a business calendar and a business day convention. In the current version, three business day conventions are implemented. The function will raise an error if an unknown business day convention is encountered. To test this function, we use the following sub procedure:

```
1 Sub TestRollDate()
2     Debug.Print RollDate(#2/15/2016#, "Follow", "NY")
3     Debug.Print RollDate(#2/15/2016#, "Preceding", "
          NY")
```

```
4      Debug.Print RollDate(#4/30/2016#, "Modified", "NY
          ")
5  End Sub
```

Executing the above sub procedure gives the following output:

```
1  2/16/2016
2  2/12/2016
3  4/29/2016
```

From the above output, we see that the function produces correct results.

In our program, payment frequencies and horizons are represented by strings that contain an integer representing a number of periods and a letter representing a period. We need a function to extract the number of periods and the period symbol from a single string, such as "1M" and "30Y." We can implement this function as follows:

```
1  Sub ExtractPeriod(ByVal sIn As String, ByRef iNum As
       Integer, ByRef sPeriod As String)
2      Dim iLen As Integer
3      Dim sNum As String
4      iLen = Len(sIn)
5
6      If iLen < 2 Then
7          Err.Raise Number:=1003, Source:="MHoliday.
              ExtractPeriod", Description:="Invalid
              period string"
8      End If
9      sPeriod = Right(sIn, 1)
10     sNum = Left(sIn, iLen - 1)
11     If IsNumeric(sNum) Then
12         iNum = CInt(sNum)
13     Else
14         Err.Raise Number:=1003, Source:="MHoliday.
              ExtractPeriod", Description:="Invalid
              period string"
15     End If
16 End Sub
```

The function ExtractPeriod has three arguments. The first argument is a ByVal argument. The last two arguments are ByRef arguments used to store the extracted number of periods and the extracted period symbol, respectively. Since the last two arguments are ByRef arguments, the procedure can change their values. To test this function, we use the following sub procedure:

```
1  Sub TestExtractPeriod()
2      Dim iNum As Integer
3      Dim sPeriod As String
4
5      ExtractPeriod "2M", iNum, sPeriod
6      Debug.Print iNum & " --- " & sPeriod
7
8      ExtractPeriod "30Y", iNum, sPeriod
9      Debug.Print iNum & " --- " & sPeriod
10 End Sub
```

Executing the above code gives

```
1  2 --- M
2  30 --- Y
```

The output shows that the function works correctly.

Now, we are ready to implement the main function for generating payment schedules. We implement this function as follows:

```
1  Function GenerateSchedule(ByVal datSet As Date, ByVal
       sFreq As String, ByVal sHorizon As String, ByVal
       sCalendar As String, ByVal sBDC As String) As
       Variant
2      Dim iFreqNum As Integer
3      Dim sFreqPeriod As String
4      Dim iHorizonNum As Integer
5      Dim sHorizonPeriod As String
6      Dim lCount As Long
7      Dim sPeriod As String
8
9      Dim vDates() As Date
10     Dim lInd As Long
11
12     ' obtain and validate input
13     If Not IsBusinessDay(datSet, sCalendar) Then
14         Err.Raise Number:=1002, Source:="MHolidy.
               GenerateSchedule", Description:="The
               settlement date is not a business day."
15     End If
16     ExtractPeriod sIn:=sFreq, iNum:=iFreqNum, sPeriod
           :=sFreqPeriod
```

```
17   ExtractPeriod sIn:=sHorizon, iNum:=iHorizonNum,
        sPeriod:=sHorizonPeriod
18
19   sPeriod = sFreqPeriod & sHorizonPeriod
20
21   If StrComp(sPeriod, "MM", vbTextCompare) = 0 Or _
22      StrComp(sPeriod, "YY", vbTextCompare) = 0 Then
23         lCount = iHorizonNum \ iFreqNum
24   ElseIf StrComp(sPeriod, "MY", vbTextCompare) = 0
        Then
25         lCount = iHorizonNum * 12 \ iFreqNum
26   ElseIf StrComp(sPeriod, "YM", vbTextCompare) = 0
        Then
27         lCount = iHorizonNum \ (iFreqNum * 12)
28   Else
29       Err.Raise Number:=1004, Source:="MHoliday.
            GenerateSchedule", Description:="Unknown
            period combination."
30   End If
31
32   If lCount <= 0 Then
33       Err.Raise Number:=1005, Source:="MHoliday.
            GenerateSchedule", Description:="Invalid
            number of periods."
34   End If
35
36   ReDim vDates(0 To lCount)
37
38   For lInd = 0 To lCount
39       vDates(lInd) = RollDate(OffsetDate(datSet,
            lInd * iFreqNum, sFreqPeriod), sBDC,
            sCalendar)
40   Next lInd
41
42   GenerateSchedule = vDates
43 End Function
```

The function GenerateSchedule takes five arguments: the settlement date, the payment frequency, the horizon of the last payment, the business calendar, and the business day count convention. This function generates an array of dates according to the given inputs and returns the array as a variant. In this function, we only considered four combinations of the payment frequency and the payment horizon. Since this is a main function, we will use the interface to test the function.

7.6.4   MInterface Module

The VBA code contained in the module MInterface is responsible for passing the inputs to the main function and writing the outputs from the main function to the worksheet.

Before we implement the macro for the button, let us define some names for the inputs. By doing so, we do not hardcode the addresses of input cells. Table 7.3 shows a list of range names we defined. On a Windows® machine, we can click "Name Manager" in the "FORMULAS" tab to see the names in the name manager window.

Table 7.3: Range Names Used by the Payment Schedule Generator

Name	Refers To
BusinessDayConvention	=Schedule!$E$6
Calendar	=Schedule!$E$5
DayCountConvention	=Schedule!$E$7
Frequency	=Schedule!$E$3
Horizon	=Schedule!$E$4
SetDate	=Schedule!$E$2

Then, we implement macro for the button as follows:

```
1  Sub Button1_Click()
2      Dim datSet As Date
3      Dim sFreq As String
4      Dim sHorizon As String
5      Dim sCalendar As String
6      Dim sBDC As String
7      Dim sDCC As String
8      Dim vSchedule As Variant
9
10     Dim lInd As Long
11     Dim lCount As Long
12
13     If IsDate(Range("SetDate").Value) Then
14         datSet = Range("SetDate").Value
15     Else
16         MsgBox "Please input a valid settlement date.
             "
17         Exit Sub
18     End If
19
```

```
20    sFreq = Range("Frequency").Value
21    sHorizon = Range("Horizon").Value
22    sCalendar = Range("Calendar").Value
23    sBDC = Range("BusinessDayConvention").Value
24    sDCC = Range("DayCountConvention").Value
25
26    vSchedule = GenerateSchedule(datSet:=datSet,
          sFreq:=sFreq, sHorizon:=sHorizon, sCalendar:=
          sCalendar, sBDC:=sBDC)
27
28    Sheet1.Range(Cells(3, 1), Cells(3, 1).End(xlDown)
          ).ClearContents
29    For lInd = LBound(vSchedule) To UBound(vSchedule)
30        Sheet1.Cells(lInd + 3, 1).Value = vSchedule(
              lInd)
31    Next lInd
32
33    Sheet1.Range(Cells(4, 2), Cells(4, 2).End(xlDown)
          ).ClearContents
34    lCount = Sheet1.Cells(3, 1).End(xlDown).Row - 2
35    With Sheet1
36        For lInd = 1 To lCount - 1
37            .Cells(lInd + 3, 2).Value = TFrac(.Cells(
                  lInd + 3, 1).Value, .Cells(lInd + 2,
                  1).Value, sDCC)
38        Next lInd
39    End With
40 End Sub
```

When the button is clicked, the sub procedure Button1_Click will run. This sub procedure is an entry point of our program. Let us look at this sub procedure. In Lines 2–9, we declare some variables. In Lines 13–18, we make sure the value in the cell named "SetDate" is a date. In Lines 20–24, we extract the inputs from relevant named cells. In Line 26, we call the GenerateSchedule function, which is defined in the module MSchedule. In Lines 28 and 33, we clear the content in the output area. In Lines 29–31, we write the schedule of dates to the output area. In Lines 35–39, we calculate the year fractions and write them to the output area.

To test the program, we use the inputs as shown in Figure 7.1. If we click the button, we see that outputs as shown in Figure 7.1.

Exercise 7.4. Write a VBA function with two arguments that returns the holiday date and the observed holiday date of a specified holiday in a specified year. The VBA function should have the following specification:

```
1 Function GetHolidayNY(ByVal lY As Long, ByVal
     sHoliday As String) As Variant
```

The argument sHoliday can take the following values:

Value	Holiday
"NewYear"	New Year's Day
"MLK"	Martin Luther King Jr. Day
"President"	Present's Day
"Memorial"	Memorial Day
"Independence"	Independence Day
"Labor"	Labor Day
"Columbus"	Columbus Day
"Veterans"	Veterans Day
"Thanksgiving"	Thanksgiving Day
"Christmas"	Christmas Day

Exercise 7.5. Create the interface shown in Figure 7.2 and write a sub procedure for the button so that clicking the button will generate the US holiday dates and the observed holiday dates for any given year. When the year is 2017, the output should be identical to what shown in Figure 7.2.

Exercise 7.6. The date of the Easter Sunday in a year Y is a day in March of the year Y. Suppose that March S, Y is the date of Easter Sunday in year Y. Then the day S is determined as follows:

(a) $N = 7 - \bmod(Y + Y\backslash 4 - Y\backslash 100 + Y\backslash 400 - 1, 7)$

(b) $G = 1 + \bmod(Y, 19)$

(c) $E = \bmod(11 \times G - 10, 30)$

(d) $H = Y\backslash 100$

(e) $SOL = H - H\backslash 4 - 12$

(f) $LUN = (H - 15 - (H - 17)\backslash 25)\backslash 3$

(g) $V = E\backslash 24 - E\backslash 25 + (G\backslash 12) \times (E\backslash 25 - E\backslash 26)$

(h) $E = \bmod(11G - 10, 30) - \bmod(SOL - LUN, 30) + V$

(i) $R = 45 - E$ if $E < 24$; $R = 75 - E$ if $E \geq 24$

(j) $C = 1 + \bmod(R + 2, 7)$

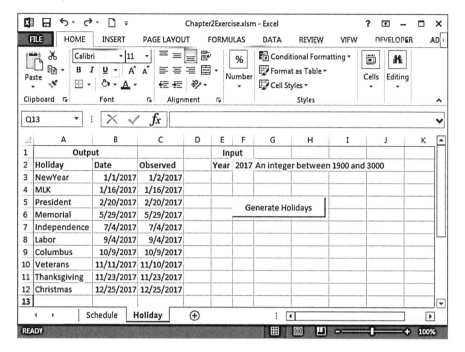

Figure 7.2: Interface of a holiday generator

(k) $S = R + \mod(7 + N - C, 7)$,

where \ denotes integer division. Write a VBA function to calculate the date of Easter Sunday in a given year. The function should have the following specification:

```
Function EasterSunday(ByVal lY As Long) As Date
```

Use your function to calculate Easter Sunday from 2010 to 2020. If your function is implemented correctly, the dates should match the dates given in Table 7.4.

7.7 Summary

Generating a schedule of payment dates is usually the first step of valuing an interest rate derivative, such as swaps. In this chapter, we introduced how

Table 7.4: Dates of Easter Sunday from 2010 to 2020

Year	Easter Sunday
2010	4/4/2010
2011	4/24/2011
2012	4/8/2012
2013	3/31/2013
2014	4/20/2014
2015	4/5/2015
2016	3/27/2016
2017	4/16/2017
2018	4/1/2018
2019	4/21/2019
2020	4/12/2020

to generate a schedule of payment dates according to a business calendar and a business day convention. We also introduced how to calculate the year fractions between two dates according to a day count convention.

We only implemented the business calendar of the United States in this chapter. However, readers can implement other business calendars in a similar way. For information on calendars, readers are referred to Richards (2012). The algorithms for interconverting Gregorian dates and Julian day numbers, and the algorithm for calculating dates of Easter Sundays are obtained from Richards (2012).

8

Bootstrapping Yield Curves

In this chapter, we implement a simple bootstrapper than can be used to construct a yield curve from a series of swap rates. Figure 8.1 shows the interface of the bootstrapper.

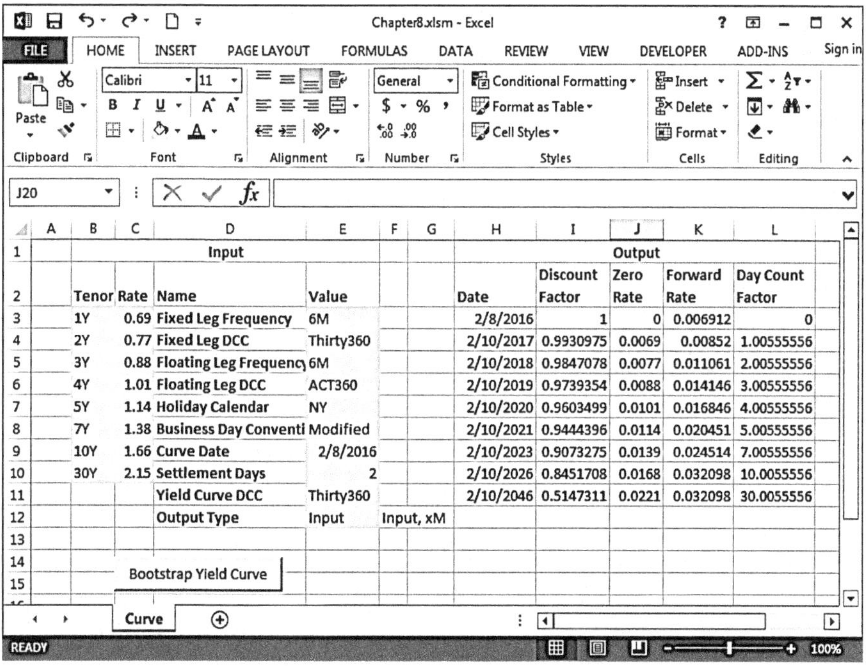

Figure 8.1: Interface of the yield curve bootstrapper.

The inputs of the bootstrapper include swap rates of various tenors (i.e., maturities), specifications of the swaps, the yield curve day count convention, and the output type. The output includes the dates, discount factors, zero rates, forward rates, and day count factors. The yield curve generator also uses two named ranges, which are given in Table 8.1.

Table 8.1: Range Names Used by the Yield Curve Bootstrapper

Name	Refers To
Param	=Curve!$E$3:$E$12
Rate	=Curve!$B$3:$C$10

8.1 Introduction

The yield curve, also known as the term structure of interest rates, is the curve that shows the relation between the interest rate and the time to maturity. For example, Table 8.2 gives the interest rates on different times to maturities (or tenors). If we plot the interest rates against the tenors, we see the yield curve as shown in Figure 8.2.

Table 8.2: Interest Rates of Various Tenors

Tenor	Rate
1 Year	0.69%
2 Year	0.77%
3 Year	0.88%
4 Year	1.01%
5 Year	1.14%
7 Year	1.39%
10 Year	1.68%
30 Year	2.21%

The interest rates shown in Table 8.2 and Figure 8.2 are continuously compounded spot rates or zero-coupon rates. The continuously compounded spot rate with maturity T is defined as

$$R(T) = -\frac{\ln P(T)}{\tau(0, T)}, \tag{8.1}$$

where $P(T)$ is the price at time 0 of a zero-coupon bond, and maturity T and $\tau(0, T)$ is the year fraction between time 0 and time T according to some day count convention (see Section 7.4). A zero-coupon bond with maturity T is a contract that guarantees the holder a cash payment of one unit at time T.

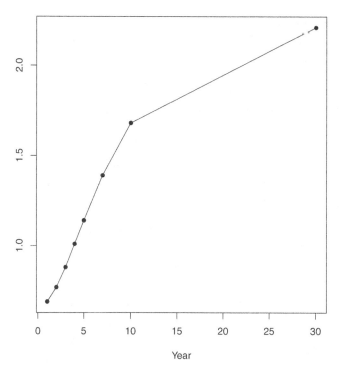

Figure 8.2: Yield curve

Let $0 \leq T < S$. Then, the forward rate for the period $[T, S]$ is defined as

$$R(T, S) = \frac{\ln P(T) - \ln P(S)}{\tau(T, S)}, \qquad (8.2)$$

where $\tau(T, S)$ is the year fraction between time T and time S according to some day count convention. The simply compounded forward rate for the period $[T, S]$ is defined as

$$F(T, S) = \frac{1}{\tau(T, S)}\left(\frac{P(T)}{P(S)} - 1\right). \qquad (8.3)$$

The discount factor is the factor that is used to multiply a future cash flow in order to obtain the present value. The discount factor used to obtained the present value of a cash flow at time T is defined as

$$DF(T) = \exp(-\tau(0, T)R(T)) = P(T). \qquad (8.4)$$

8.2 Interpolation

Interest rate products in the market have only a few different maturities. In order to construct a yield curve, we need to know the interest rates at all tenors within a time horizon (e.g., 30 years). Interpolation is used to estimate the interest rates at other maturities. Interpolation is also connected to the bootstrapping method described in Section 8.3.

There are several interpolation methods (Hagan and West, 2006, 2008): linear on rates, linear on the log of rates, linear on discount factors, linear on the log of discount factors, piecewise linear forward, and splines. The linear on the log of discount factors method, also known as the raw interpolation method, is stable and straightforward to implement. In this program, we use the raw interpolation method.

Let $T_i < T_{i+1}$ be two maturities of which we know the discount factors. Let $T \in [T_i, T_{i+1}]$. Then, in the raw interpolation method, the discount factor $DF(T)$ at time T is interpolated from the discount factors $DF(T_i)$ and $DF(T_{i+1})$ as follows:

$$\ln DF(T) = \frac{\tau(T, T_{i+1})\ln DF(T_i) + \tau(T_i, T)\ln DF(T_{i+1})}{\tau(T_i, T_{i+1})}. \tag{8.5}$$

By Equation (8.4), the above equation is equivalent to

$$R(T) = \frac{\tau(T, T_{i+1})T_i R(T_i) + \tau(T_i, T)T_{i+1} R(T_{i+1})}{\tau(T_i, T_{i+1})T}. \tag{8.6}$$

For other interpolation methods, readers are referred to (Hagan and West, 2006) and (Hagan and West, 2008).

8.3 Bootstrapping Yield Curves

In this section, we show how to construct a yield curve from a series of swap rates using the bootstrapping method. One advantage of using swap rates in the construction of yield curves is that we observe swap rates from a wide number of maturities. For example, Table 8.3 gives the US swap rates for various maturities on February 8, 2016.

To construct a yield curve from the swap rates for the maturities shown in Table 8.3, we bootstrap the discount factors for the periods corresponding

Table 8.3: US Swap Rates for Various Maturities on February 8, 2016. These rates were obtained from `http://www.federalreserve.gov`

Tenor	Rate
1Y	0.69%
2Y	0.77%
3Y	0.88%
4Y	1.01%
5Y	1.14%
7Y	1.38%
10Y	1.66%
30Y	2.15%

to these maturities. Discount factors for other periods can be interpolated from those discount factors.

To describe the bootstrapping method, let us introduce how interest rate swaps are priced. As we mentioned in Chapter 7, an interest rate swap (IRS) is a financial derivative in which two parties agree to exchange interest rate cash flows, based on a specified notional amount. A typical IRS consists of two legs of cash flows: one leg of incoming cash flows, and one leg of outgoing cash flows. Usually, the cash flows of one leg are calculated according to a fixed rate, and the cash flows of another leg are calculated according to a floating rate. In such cases, we say that the swap has a fixed leg and a floating leg. Some swaps consist of two floating legs.

The swap rates given in Table 8.3 are the rates of swaps that consist of a fixed leg and a floating leg. In addition, the swap rates are determined in such a way that the corresponding swaps have initial values of zero. In other words, the swap rate is determined such that the present value of the fixed leg cash flows is equal to that of the floating leg cash flows.

Let T_0 be the settlement date, and $T_1, T_2, \ldots, T_n$ be the payment dates of an IRS. In the US, the settlement date is two business days after the trade date. An IRS usually has semiannual payments. If a swap matures in one year, then it has two cash flows: one in 6 months and one in the maturity. If a swap matures in 30 years, then it has 60 cash flows.

Suppose that the notional amount is one unit. For $i = 1, 2, \ldots, n$, the fixed cash flow at the date T_i is calculated as

$$\tau(T_{i-1}, T_i)r,$$

where $\tau(T_{i-1}, T_i)$ is the year fraction between dates T_{i-1} and T_i, and r is the swap rate. The floating cash flow at the date T_i is calculated as

$$\tau(T_{i-1}, T_i)F(T_{i-1}, T_i),$$

where $F(T_{i-1}, T_i)$ is the simply compounded forward rate for the period $[T_{i-1}, T_i]$ (see Equation (8.3)). The swap rate is determined to satisfy the following equation:

$$\sum_{i=1}^{n} \tau(T_{i-1}, T_i)rDF(T_i) = \sum_{i=1}^{n} \tau(T_{i-1}, T_i)F(T_{i-1}, T_i)DF(T_i), \qquad (8.7)$$

where $DF(T_i) = P(T_i)$ is the discount factor for the period $[0, T_i]$. Since

$$
\begin{aligned}
\sum_{i=1}^{n} \tau(T_{i-1}, T_i)F(T_{i-1}, T_i)DF(T_i) &= \sum_{i=1}^{n} \left(\frac{P(T_{i-1})}{P_{T_i}} - 1 \right) P(T_i) \\
&= \sum_{i=1}^{n} [P(T_{i-1}) - P(T_i)] \\
&= P(T_0) - P(T_n),
\end{aligned}
$$

Equation (8.7) is reduced to

$$r \sum_{i=1}^{n} \tau(T_{i-1}, T_i)DF(T_i) = DF(T_0) - DF(T_n). \qquad (8.8)$$

According to Equation (8.8), we can solve the swap rate r given a yield curve, because we can obtain the discount factors $DF(T_0), DF(T_1), \ldots, DF(T_n)$ from the yield curve. The bootstrapping procedure reverses this process, that is, we solve the discount factors $DF(T_0), DF(T_1), \ldots, DF(T_n)$ given the swap r observed in the market. However, there are many unknown discount factors and only one equation (i.e., Equation (8.8)). How can we solve the discount factors based on one equation? To solve this problem, we just treat $DF(T_n)$ as an unknown in the equation and interpolate other unknown discount factors from $DF(T_n)$ and another known discount factor, which can be 1 (i.e., the discount factor at time 0) or a discount factor bootstrapped from a swap rate for a shorter maturity.

Let T_0 be the settlement date. Let T_i denote the payment date after $6i$ months, for $i = 1, 2, \ldots, 60$. In particular, the bootstrapping procedure works as follows:

(a) Bootstrap the discount factor $DF(T_2)$ using the 1Y swap rate. In this step, we need to interpolate the discount factors at T_0 (settlement date) and T_1 (6 months) from 1 (the discount factor at time 0) and $DF(T_2)$. We use a method described in Section 8.4 to solve $DF(T_2)$.

(b) Bootstrap the discount factor $DF(T_4)$ using the 2Y swap rate. In this step, we already know the discount factor $DF(T_0), DF(T_1)$, and $DF(T_2)$ from the previous step. We only need to interpolate the discount factor $DF(T_3)$ from $DF(T_2)$ and $DF(T_4)$. We can use the same method to solve $DF(T_4)$.

(c) Repeat the above process to bootstrap discount factors $DF(T_6)$, $DF(T_8)$, $DF(T_{10})$, $DF(T_{14})$, $DF(T_{20})$, and $DF(T_{60})$ from the swap rates for maturities 3Y, 4Y, 5Y, 7Y, 10Y, and 30Y.

8.4 Finding Roots of an Equation

Bootstrapping discount factors from swap rates requires solving nonlinear equations. For example, when we bootstrap the discount factor $DF(T_2)$ from the 1Y swap rate, we need to solve the following equation:

$$r\tau(T_0, T_1)DF(T_1) + r\tau(T_1, T_2)DF(T_2) = DF(T_0) - DF(T_2),$$

where $DF(T_0)$ and $DF(T_1)$ are interpolated from 1 and $DF(T_2)$. If we treat $DF(T_2)$ as an unknown x in the equation, then the above equation can be written as

$$r\tau(T_0, T_1)g(T_1, x) + r\tau(T_1, T_2)x = g(T_0, x) - x,$$

or

$$r\tau(T_0, T_1)g(T_1, x) + r\tau(T_1, T_2)x - g(T_0, x) + x = 0,$$

where $g(\cdot, \cdot)$ denotes the interpolation function. Solving the above equation for x gives the discount factor $DF(T_2)$.

In this section, we introduce Newton's method, also known as the Newton–Raphson method, to find a root of the following equation:

$$f(x) = 0.$$

Newton's method is an iterative method (Press et al., 2002, Chapter 9). Let x_0 and x_1 be two initial guesses of the root of the above equation. Then, Newton's method proceeds as follows:

$$x_{i+1} = x_i - (x_i - x_{i-1})\frac{f(x_i)}{f(x_i) - f(x_{i-1})}, \quad i = 1, 2, \dots. \quad (8.9)$$

We stop the iterative process until some condition is met. For example, we can stop the iterative process when $|x_{i+1} - x_i| < 1e - 10$.

8.5   Implementation

As we can see from Figure 8.1, the interface of the bootstrapping tool contains inputs, outputs, and one button. When we click the button, the yield curve will be bootstrapped from the swap rates given in the input. The Visual Basic for Applications (VBA) code behind the interface is organized into five modules: MDate, MHoliday, MSchedule, MCurve, and MInterface. The first three modules, MDate, MHoliday, and MSchedule, are imported from the payment schedule generator implemented in Chapter 7. We will implement the modules MCurve and MInterface in this section.

8.5.1   MCurve Module

The module MCurve contains VBA code that implements the bootstrapping technique described in previous sections of this chapter. The declaration section of this module contains the following code:

```
 1 Option Explicit
 2
 3 ' input
 4 Dim aTenor() As String
 5 Dim aRate() As Double
 6 Dim sFixFreq As String
 7 Dim sFixDCC As String
 8 Dim sFltFreq As String
 9 Dim sFltDCC As String
10 Dim sCalendar As String
11 Dim sBDC As String
12 Dim datCurveDate As Date
13 Dim datSet As Date
14 Dim iSetDays As Integer
15 Dim sYCDCC As String
16 Dim sOutputType As String
17
18 ' yield curve
19 Dim aDate() As Date    ' tenor of the rates
20 Dim aDF() As Double    ' discount factor
```

In the above code, we defined quite a few module-level variables that are used to store the input information and the output yield curve. The swap rates and their tenors are saved in the arrays aRate and aTenor, respectively. The resulting yield curve is stored in the arrays aDate and aDF, respectively.

The array aDate contains the maturity dates of the swap rates, and the array aDF contains the discount factors at these maturity dates.

The first procedure of this module is the sub procedure Initialize, which is used to initialize the module-level variables based on the input information. It also calculates the settlement date based on the number of settlement days and the business calendar. We implement this sub procedure as follows:

```vba
Sub Initialize(ByRef rngRate As Range, ByRef rngParam
    As Range)
    Dim iNumRates As Integer
    Dim i As Integer
    iNumRates = rngRate.Rows.Count

    ReDim aTenor(1 To iNumRates)
    ReDim aRate(1 To iNumRates)

    For i = 1 To iNumRates
        aTenor(i) = rngRate.Cells(i, 1).Value
        aRate(i) = rngRate.Cells(i, 2).Value / 100
    Next i

    sFixFreq = rngParam.Cells(1, 1).Value
    sFixDCC = rngParam.Cells(2, 1).Value
    sFltFreq = rngParam.Cells(3, 1).Value
    sFltDCC = rngParam.Cells(4, 1).Value
    sCalendar = rngParam.Cells(5, 1).Value
    sBDC = rngParam.Cells(6, 1).Value
    datCurveDate = rngParam.Cells(7, 1).Value
    iSetDays = rngParam.Cells(8, 1).Value
    sYCDCC = rngParam.Cells(9, 1).Value
    sOutputType = rngParam.Cells(10, 1).Value

    datSet = datCurveDate
    For i = 1 To iSetDays
        datSet = RollDate(AddDays(datSet, 1), "Follow
            ", sCalendar)
    Next i

    ReDim aDate(0 To 0)
    ReDim aDF(0 To 0)
    aDate(0) = datCurveDate
    aDF(0) = 1
End Sub
```

In this sub procedure, the settlement date is calculated, and the yield curve is initialized. The output yield curve is stored in two array-type variables: aDate and aDF. At the beginning, the variable aDate is initialized to contain the curve date, and the variable aDF is initialized to contain 1, which is the discount factor at the curve date.

For testing purposes, we also create the following sub procedure:

```
1  Sub PrintInfo()
2      Dim i As Integer
3      For i = LBound(aTenor) To UBound(aTenor)
4          Debug.Print "Swap Rate " & i & ": " & aTenor(
               i) & " -- " & aRate(i)
5      Next i
6
7      Debug.Print "Fixed Leg: " & sFixFreq & " -- " &
               sFixDCC
8      Debug.Print "Floating Leg: " & sFltFreq & " -- "
               & sFltDCC
9      Debug.Print "Calendar: " & sCalendar
10     Debug.Print "Business Day Convention: " & sBDC
11     Debug.Print "Curve Date: " & datCurveDate
12     Debug.Print "Settlement Days: " & iSetDays
13     Debug.Print "Yield Curve Day Count Convention: "
               & sYCDCC
14     Debug.Print "Settlement Date: " & datSet
15
16     For i = LBound(aDate) To UBound(aDate)
17         Debug.Print "Discount Factor " & i & ": " &
               aDate(i) & " -- " & aDF(i)
18     Next i
19
20 End Sub
```

This sub procedure just displays the content of the module-level variables in the Immediate Window.

To test the sub procedure Initialize, we use the following code:

```
1  Sub TestInitialize()
2      Initialize rngRate:=Range("Rate"), rngParam:=
           Range("Param")
3
4      PrintInfo
5  End Sub
```

Executing the above sub procedure produces the following output:

```
 1 Swap Rate 1: 1Y -- 0.0069
 2 Swap Rate 2: 2Y -- 0.0077
 3 Swap Rate 3: 3Y -- 0.0088
 4 Swap Rate 4: 4Y -- 0.0101
 5 Swap Rate 5: 5Y -- 0.0114
 6 Swap Rate 6: 7Y -- 0.0138
 7 Swap Rate 7: 10Y -- 0.0166
 8 Swap Rate 8: 30Y -- 0.0215
 9 Fixed Leg: 6M -- Thirty360
10 Floating Leg: 6M -- ACT360
11 Calendar: NY
12 Business Day Convention: Modified
13 Curve Date: 2/8/2016
14 Settlement Days: 2
15 Yield Curve Day Count Convention: Thirty360
16 Settlement Date: 2/10/2016
17 Discount Factor 0: 2/8/2016 -- 1
```

If we compare the above output to the inputs shown in Figure 8.1, we see that the sub procedure Initialize works correctly.

We implement the raw interpolation method as follows:

```
 1 Function LogLinear(ByVal datD As Date) As Double
 2     If datD < datCurveDate Then
 3         Err.Raise Number:=2001, Source:="MCurve.
               LogLinear", Description:="Date is ealier
               than the curve date."
 4     End If
 5     Dim i As Integer
 6     Dim iL As Integer
 7     Dim iH As Integer
 8     iL = LBound(aDate)
 9     iH = UBound(aDate)
10     For i = iL To iH
11         If datD = aDate(i) Then
12             LogLinear = aDF(i)
13             Exit Function
14         End If
15     Next i
16
17     Dim dat1 As Date
18     Dim dat2 As Date
19     Dim dDF1 As Double
```

```
20      Dim dDF2 As Double
21      Dim bFound As Boolean
22      bFound = False
23      For i = iL To iH - 1
24          If datD > aDate(i) And datD < aDate(i + 1)
                Then
25              dat1 = aDate(i)
26              dat2 = aDate(i + 1)
27              dDF1 = aDF(i)
28              dDF2 = aDF(i + 1)
29              bFound = True
30          End If
31      Next i
32
33      If Not bFound Then
34          dat1 = aDate(iH - 1)
35          dat2 = aDate(iH)
36          dDF1 = aDF(iH - 1)
37          dDF2 = aDF(iH)
38      End If
39
40      Dim dDCF1 As Double
41      Dim dDCF2 As Double
42      Dim dDCF As Double
43      dDCF1 = TFrac(datCurveDate, dat1, sYCDCC)
44      dDCF2 = TFrac(datCurveDate, dat2, sYCDCC)
45      dDCF = TFrac(datCurveDate, datD, sYCDCC)
46
47      Dim dTemp As Double
48      dTemp = WorksheetFunction.Ln(dDF2) * (dDCF -
            dDCF1) + _
49          WorksheetFunction.Ln(dDF1) * (dDCF2 - dDCF)
50      LogLinear = Exp(dTemp / (dDCF2 - dDCF1))
51 End Function
```

The function procedure LogLinear has one argument, which is a date. The function will raise an error if the input date is older than the curve date. If the input date is equal to one of the dates stored in the variable aDate, the function returns the discount factor corresponding to this date. If the date is not equal to any dates stored in the variable aDate, the function returns the discount factor interpolated by the raw interpolation method.

To test the function LogLinear, we use the following sub procedure:

```
1 Sub TestLogLinear()
```

```
 2      ReDim aDate(0 To 2)
 3      ReDim aDF(0 To 2)
 4
 5      aDate(0) = #4/5/2016#
 6      aDate(1) = #4/5/2017#
 7      aDate(2) = #4/5/2026#
 8      aDF(0) = 1
 9      aDF(1) = 0.95
10      aDF(2) = 0.7
11
12      sYCDCC = "Thirty360"
13
14      Debug.Print LogLinear(#4/5/2016#)
15      Debug.Print LogLinear(#10/5/2016#)
16      Debug.Print LogLinear(#4/5/2041#)
17 End Sub
```

Executing the above code gives the following output:

```
 1   1
 2   0.974679434480896
 3   0.420780331300377
```

The output indicates that the function works appropriately.
We calculate the present value of an IRS as follows:

```
 1      Dim iNum As Integer
 2      Dim sPeriod As String
 3      Dim i As Integer
 4      Dim dDF As Double
 5
 6      Dim vFixDates As Variant
 7      Dim dFixPV As Double
 8      ExtractPeriod sIn:=sTenor, iNum:=iNum, sPeriod:=
          sPeriod
 9
10      ' calculate fixed leg cash flows and present
          value
11      vFixDates = GenerateSchedule(datSet, sFixFreq,
          sTenor, sCalendar, sBDC)
12      For i = LBound(vFixDates) + 1 To UBound(vFixDates
          )
13          dDF = LogLinear(vFixDates(i))
14          dFixPV = dFixPV + TFrac(vFixDates(i - 1),
              vFixDates(i), sFixDCC) * dRate * dDF
```

```
15        Next i
16
17        ' calculate floating leg cash flows and present
              value
18        Dim vFltDates As Variant
19        Dim dFltPV As Double
20        Dim dFR As Double
21        Dim dT As Double
22        ' calculate fixed leg cash flows and present
              value
23        vFltDates = GenerateSchedule(datSet, sFltFreq,
              sTenor, sCalendar, sBDC)
24        For i = LBound(vFltDates) + 1 To UBound(vFltDates
              )
25            dDF = LogLinear(vFltDates(i))
26            dT = TFrac(vFltDates(i - 1), vFltDates(i),
                  sFltDCC)
27            dFR = (LogLinear(vFltDates(i - 1)) /
                  LogLinear(vFltDates(i)) - 1) / dT
28            dFltPV = dFltPV + dT * dFR * dDF
29        Next i
30
31        PvSwap = dFixPV - dFltPV
32  End Function
```

The function procedure PvSwap has two arguments: the tenor and the rate of the swap. This function uses some module-level variables, such as these used to specify the payment frequency and the business calendar. It first calculates the present value of the fixed leg, and then calculates the present value of the floating leg.

To test the function PvSwap, we use the following sub procedure:

```
1  Sub TestPvSwap()
2      Initialize rngRate:=Range("Rate"), rngParam:=
           Range("Param")
3
4      ReDim Preserve aDate(0 To 1)
5      ReDim Preserve aDF(0 To 1)
6
7      aDate(1) = OffsetDate(datSet, 1, "Y")
8      aDF(1) = 0.95
9
10     Debug.Print PvSwap(aTenor(1), aRate(1))
11 End Sub
```

If we use the inputs shown in Figure 8.1, then executing the above code gives

```
1 | -4.30769841682971E-02
```

The output shows that the present value of the fixed leg is lower than that of the floating leg.

We implement Newton's method as follows:

```
 1 | Sub SolveRate(ByVal iInd As Integer)
 2 |         ReDim Preserve aDate(0 To iInd)
 3 |         ReDim Preserve aDF(0 To iInd)
 4 |
 5 |         Dim iNum As Integer
 6 |         Dim sPeriod As String
 7 |
 8 |         ExtractPeriod sIn:=aTenor(iInd), iNum:=iNum,
          sPeriod:=sPeriod
 9 |         aDate(iInd) = OffsetDate(datSet, iNum,
          sPeriod)
10 |
11 |         Dim dx As Double
12 |         Dim dx0 As Double
13 |         Dim dx1 As Double
14 |         Dim df As Double
15 |         Dim df0 As Double
16 |         Dim df1 As Double
17 |
18 |         dx0 = Exp(-aRate(iInd) * TFrac(datCurveDate,
          aDate(iInd), sYCDCC))
19 |         aDF(iInd) = dx0
20 |         df0 = PvSwap(aTenor(iInd), aRate(iInd))
21 |         dx1 = dx0 + 0.001
22 |         aDF(iInd) = dx1
23 |         df1 = PvSwap(aTenor(iInd), aRate(iInd))
24 |         Do While Abs(dx1 - dx0) > 0.0000000001
25 |             dx = dx0 - df0 * (dx1 - dx0) / (df1 - df0
                )
26 |             aDF(iInd) = dx
27 |             df = PvSwap(aTenor(iInd), aRate(iInd))
28 |
29 |             dx0 = dx1
30 |             df0 = df1
31 |             dx1 = dx
32 |             df1 = df
```

```
33          Loop
34
35          aDF(iInd) = dx
36 End Sub
```

The sub procedure `SolveRate` has only one argument, which is an integer that represents the index of the input swap rates. This sub procedure also modifies the variables used to store the yield curve. At the beginning of this sub procedure, the variables aDate and aDF are redimensioned to contain one more value. The existing values are preserved.

To test the sub procedure `SolveRate`, we use the following sub procedure:

```
1 Sub TestSolveRate()
2      Initialize rngRate:=Range("Rate"), rngParam:=
         Range("Param")
3
4      SolveRate iInd:=1
5
6      Debug.Print aDate(1) & " -- " & aDF(1)
7 End Sub
```

If we use the inputs shown in Figure 8.1, then executing the above code gives

```
1 2/10/2017 -- 0.993097540002669
```

The output shows that the bootstrapped discount factor for 1 year is about 0.9931.

Now, we can implement a sub procedure to bootstrap all the discount factors as follows:

```
1 Sub BuildCurve()
2      Dim iL As Integer
3      Dim iH As Integer
4      Dim i As Integer
5      iL = LBound(aTenor)
6      iH = UBound(aTenor)
7
8      For i = iL To iH
9          SolveRate iInd:=i
10     Next i
11 End Sub
```

The sub procedure `BuildCurve` just loops through the input swap rates and calls the sub procedure `SolveRate` to bootstrap the discount factors. To test this sub procedure, we use the following sub procedure:

```
1  Sub TestBuildCurve()
2      Dim i As Integer
3      Initialize rngRate:=Range("Rate"), rngParam:=
           Range("Param")
4
5      BuildCurve
6      PrintInfo
7  End Sub
```

Executing the above code gives the following output:

```
1  Swap Rate 1: 1Y -- 0.0069
2  Swap Rate 2: 2Y -- 0.0077
3  Swap Rate 3: 3Y -- 0.0088
4  Swap Rate 4: 4Y -- 0.0101
5  Swap Rate 5: 5Y -- 0.0114
6  Swap Rate 6: 7Y -- 0.0138
7  Swap Rate 7: 10Y -- 0.0166
8  Swap Rate 8: 30Y -- 0.0215
9  Fixed Leg: 6M -- Thirty360
10 Floating Leg: 6M -- ACT360
11 Calendar: NY
12 Business Day Convention: Modified
13 Curve Date: 2/8/2016
14 Settlement Days: 2
15 Yield Curve Day Count Convention: Thirty360
16 Settlement Date: 2/10/2016
17 Discount Factor 0: 2/8/2016 -- 1
18 Discount Factor 1: 2/10/2017 -- 0.993097540002669
19 Discount Factor 2: 2/10/2018 -- 0.984707793704918
20 Discount Factor 3: 2/10/2019 -- 0.973935374151509
21 Discount Factor 4: 2/10/2020 -- 0.960349877939723
22 Discount Factor 5: 2/10/2021 -- 0.944439556311098
23 Discount Factor 6: 2/10/2023 -- 0.907327535413443
24 Discount Factor 7: 2/10/2026 -- 0.845170756713933
25 Discount Factor 8: 2/10/2046 -- 0.514731101479042
```

From the output, we see that the yield curve was constructed.

Finally, we implement a function to create output based on a specified output type. We implement this function as follows:

```
1  Function CreateOutput() As Variant
2      Dim aOutDates() As Date
3      Dim i As Integer
4      Dim iL As Integer
5      Dim iH As Integer
6      Dim dT As Double
7      If StrComp(sOutputType, "Input", vbTextCompare) =
          0 Then
8          iL = LBound(aDate)
9          iH = UBound(aDate)
10         ReDim aOutDates(iL To iH)
11         For i = iL To iH
12             aOutDates(i) = aDate(i)
13         Next i
14     Else
15         aOutDates = GenerateSchedule(datCurveDate,
              sOutputType, "30Y", sCalendar, sBDC)
16     End If
17
18     iL = LBound(aOutDates)
19     iH = UBound(aOutDates)
20
21     ReDim aOut(iL To iH, 0 To 4) As Double
22     For i = iL To iH
23         aOut(i, 0) = CDbl(aOutDates(i))
24         aOut(i, 1) = LogLinear(aOutDates(i))
25         dT = TFrac(datCurveDate, aOutDates(i), sYCDCC
              )
26         If dT = 0 Then
27             aOut(i, 2) = 0
28         Else
29             aOut(i, 2) = -WorksheetFunction.Ln(aOut(i
                  , 1)) / dT
30         End If
31         If i < iH Then
32             aOut(i, 3) = (LogLinear(aOutDates(i)) /
                  LogLinear(aOutDates(i + 1)) - 1) /
                  TFrac(aOutDates(i), aOutDates(i + 1),
                  sYCDCC)
33         Else
34             aOut(i, 3) = aOut(i - 1, 3)
35         End If
36         aOut(i, 4) = TFrac(datCurveDate, aOutDates(i)
              , sYCDCC)
37     Next i
```

```
38      CreateOutput = aOut
39  End Function
```

The sub procedure CreateOutput calculates other relevant quantities (e.g., forward rates, zero rates, and day count factors) based on the yield curve and returns a two-dimensional array as a variant to the caller. If the output type is "Input," this sub procedure calculates these quantities only at the maturities of the input swap rates. If the output is "xM" (e.g., 1M, 3M, 12M, etc.), this sub procedure calculates these quantities at a sequence of dates generated by the given frequency. We will test this function using the interface.

8.5.2 MInterface Module

The module MInterface contains VBA code for the button. We implement this sub procedure as follows:

```
1  Sub Button1_Click()
2      Dim vRes As Variant
3      Initialize rngRate:=Range("Rate"), rngParam:=
           Range("Param")
4      BuildCurve
5      vRes = CreateOutput
6
7      Dim iRows As Integer
8      Dim i As Integer
9      Dim j As Integer
10     iRows = Sheet1.Range("H3").End(xlDown).Row
11     Sheet1.Range(Cells(3, 8), Cells(iRows, 12)).
           ClearContents
12     For i = LBound(vRes, 1) To UBound(vRes, 1)
13         For j = 0 To 4
14             Sheet1.Cells(i + 3, 8 + j).Value = vRes(i
                   , j)
15         Next j
16     Next i
17  End Sub
```

This sub procedure is responsible for preparing the input and displaying the output. The sub procedure first calls other three functions: Initialize, BuildCurve, and CreateOutput. The three functions are defined in the module MCurve. Then, the sub procedure writes the output to columns 8 to 12. If we want to rearrange the input or change the output to a different location, we only need to change the VBA code in the module `MInterface`.

To test the program, we use the inputs as shown in Figure 8.1. If we click the button, we see that outputs as shown in Figure 8.1.

Exercise 8.1. The bootstrapping procedure first bootstraps the discount factor for the 1Y swap rate, then bootstraps the discount factor for the 2Y swap rate, and so on. However, the VBA program does not sort the input swap rates in terms of tenors. Instead, the VBA program bootstraps the discount factors based on the input order of the swap rates. Write a sub procedure named `CheckInput` to verify that the input order of the swap rates is an increasing order of tenors. This sub procedure is included in the module `MCurve` and should be called after the procedure `Initialize` and before the procedure `BuildCurve`. The sub procedure `CheckInput` should raise an error if the input order of the swap rates is not an increasing order of tenors.

Exercise 8.2. The function procedure `PvSwap` in the module `MCurve` calculates the present value of the floating leg by projecting and discounting the individual cash flows. In fact, we do not have to do this. From Equation (8.8), the net present value of a swap is given by

$$r \sum_{i=1}^{n} \tau(T_{i-1}, T_i) DF(T_i) - DF(T_0) + DF(T_n).$$

Rewrite the function procedure `PvSwap` by using the above formula to calculate the net present value.

Exercise 8.3. Newton's method may not converge if the initial guesses are not good. Another method for finding a root of an equation is the bisection method. Suppose that we want to find a root of the equation $f(x) = 0$, where $f(x)$ is a continuous function of x. We first find two values $x_L < x_U$, such that $f(x_L)f(x_U) < 0$. Then we know that there is a root in the interval (x_L, x_U). We can use the bisection method to find the root. To do that, we let

$$x_M = \frac{x_L + x_H}{2}.$$

If $f(x_L)f(x_M) < 0$, then we update $x_H = x_M$. If $f(x_M)f(x_H) < 0$, then we update $x_L = x_M$. We continues updating x_L and x_H in the above way until $|x_L - x_H| < 10^{-10}$.

Rewrite the sub procedure `SolveRate` by using the bisection method described above. You can use the following initial values $x_L - \exp(-10\,r\,T)$ and $x_H = 1$, where r denotes the swap rate under consideration and T denotes its maturity.

Exercise 8.4. Modify the sub procedure `Button1_Click` to do the following: Create a chart sheet named `Plot`, if it does not exist, and plot the zero rates and the forward rates against the day count factors in the chart sheet. (Hint: you can use the macro recorder to get started.)

8.6 Summary

Bootstrapping a yield curve allows us to produce swap prices that are consistent with market prices. In this chapter, we introduced how to construct a yield curve by bootstrapping discount factors from swap rates, which cover a wide range of maturities. We also introduced an interpolation method and Newton's method for solving a nonlinear equation that is required in the bootstrapping process. For more information about interpolation methods for constructing yield curves, readers are referred to (Hagan and West, 2006, 2008). Other methods for solving nonlinear equations can be found in (Press et al., 2002; Garrett, 2015).

9

Generating Risk-Neutral Scenarios

In this chapter, we implement a simple economic scenario generator for generating risk-neutral scenarios of an equity security (e.g., a stock). The interface of this risk-neutral scenario generator is shown in Figure 9.1.

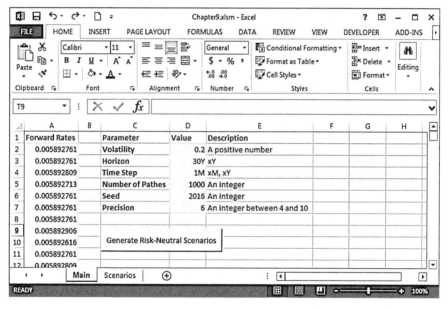

Figure 9.1: Interface of the risk-neutral scenario generator

The inputs of the scenario generator include a list of forward rates and six parameters: the volatility, the horizon of the scenarios, the time step, the number of paths, a seed, and the precision of the scenarios. The seed allows us to repeat the scenarios. In other words, if we use the same seed and the same other parameters, then we get the same set of scenarios.

9.1   Introduction

Economic scenario generators have been used by insurance companies to price insurance products (e.g., variable annuities) embedding guarantees and calculate capital adequacy. An economic scenario generator is simply a computer program that produces random sequences of market variables (e.g., equity returns) by assuming that those market variables follow some stochastic process. For a particular market variable, we call a random sequence of the market variable a path or a scenario.

There are two types of economic scenarios that are commonly used by insurance companies: risk-neutral scenarios and real-world scenarios. Risk-neutral scenarios, also referred to as arbitrage-free or market-consistent scenarios, are economic scenarios that are consistent with market prices. Risk-neutral scenarios are generally used for pricing purpose. Real-world scenarios are created based on the historical experience of market variables and are used for other purposes, such as calculating capital adequacy and profitability of a product.

9.2   Black–Scholes Model

There are numerous approaches to generate risk-neutral scenarios for an equity security. One common approach is to use the Black–Scholes model (Black and Scholes, 1973). Under the Black–Scholes model, the equity security is assumed to follow the following stochastic process (Carmona and Durrleman, 2005):

$$S_t = S_0 \exp\left[\left(\int_0^t r_s \mathrm{d}s - \frac{t}{2}\sigma^2\right) + \sigma B_t\right], \quad t \geq 0, \tag{9.1}$$

where r_s is the short rate at time s, σ is the volatility of the equity security, and $\{B_t : t \geq 0\}$ is a standard Brownian motion (Gan et al., 2014).

The stochastic process given in Equation (9.1) is a continuous-time stochastic process. Since a path is a discrete sequence, we need to discretize the continuous-time stochastic process. Let Δ be a time step in years. For example, if we want to generate monthly scenarios, then $\Delta = \frac{1}{12}$. If we want to generate scenarios at the annual time step, then $\Delta = 1$. Let m be the number of time steps of a path. For example, if we want to generate scenarios at the monthly time step for a horizon of 30 years, then $m = 360$. For $i = 0, 1, \ldots, m$, let $t_i = \Delta i$. Then, by Equation (9.1), we have

$$\frac{S_{t_i}}{S_{t_{i-1}}} = \exp\left[\left(\int_{t_{i-1}}^{t_i} r_s ds - \frac{\Delta}{2}\sigma^2\right) + \sigma(B_{t_i} - B_{t_{i-1}})\right] \tag{9.2}$$

for $i = 1, 2, \ldots, m$.

Let

$$A_i = \frac{S_{t_i}}{S_{t_{i-1}}}, \quad i = 1, 2, \ldots, m. \tag{9.3}$$

We call A_i the accumulation factor for the ith period. In other words, A_i is the growth of the equity security from time $t_{i-1} = (i-1)\Delta$ to time $t_i = i\Delta$. Then, by Equation (9.2) and Equation (9.3), we have

$$\ln A_i = \left(\int_{t_{i-1}}^{t_i} r_s ds - \frac{\Delta}{2}\sigma^2\right) + \sigma(B_{t_i} - B_{t_{i-1}}), \tag{9.4}$$

for $i = 1, 2, \ldots, m$. The logarithm of A_i, $\ln A_i$, is the continuous return of the equity security for the ith period.

Let

$$Z_i = \frac{B_{t_i} - B_{t_{i-1}}}{\sqrt{\Delta}}, \quad i = 1, 2, \ldots, m. \tag{9.5}$$

Then, by the property of Brownian motions, we know that Z_i is a standard normal random variable (i.e., a normal random variable with a mean of 0 and a standard deviation of 1), and that $Z_1, Z_2, \ldots, Z_m$ are independent.

We assume that the interest rate is a deterministic function of time, and the forward rate is constant within each period. For $i = 1, 2, \ldots, m$, let f_i be the annualized simply compounded forward rate for the period $[t_{i-1}, t_i]$. Then, by Equation (8.3), we have

$$f_i = \frac{1}{\Delta}\left(\frac{P(t_{i-1})}{P(t_i)} - 1\right), \tag{9.6}$$

where $P(t_i)$ is the price at time zero of the zero-coupon bond maturing at time t_i. Since the interest rate is deterministic rather than stochastic, we have (Gan et al., 2014, Chapter 45)

$$P(t_i) = \exp\left(-\int_0^{t_i} r_s ds\right), \quad i = 0, 1, \ldots, m. \tag{9.7}$$

Combining Equation (9.6) and Equation (9.7), we get

$$f_i = \frac{1}{\Delta}\left[\exp\left(\int_{t_{i-1}}^{t_i} r_s ds\right) - 1\right], \tag{9.8}$$

which gives

$$\int_{t_{i-1}}^{t_i} r_s ds = \ln(1 + \Delta f_i). \tag{9.9}$$

Now, plugging Equation (9.5) and Equation (9.9) into Equation (9.4), we get

$$\ln A_i = \ln(1 + \Delta f_i) - \frac{\Delta}{2}\sigma^2 + \sigma\sqrt{\Delta}Z_i \tag{9.10}$$

for $i = 1, 2, \ldots, m$. We use Equation (9.10) to generate risk-neutral scenarios.

Let n be the number of paths or scenarios we want to generate. Then, the scenarios can be represented by an $n \times m$ matrix defined as

$$\begin{pmatrix} A_{11} & A_{12} & \cdots & A_{1m} \\ A_{21} & A_{22} & \cdots & A_{2m} \\ \vdots & \vdots & \ddots & \vdots \\ A_{n1} & A_{n2} & \cdots & A_{nm} \end{pmatrix}, \tag{9.11}$$

where A_{ij}, called the accumulation factor for the jth period of the ith scenario, is generated as

$$\ln A_{ij} = \ln(1 + \Delta f_j) - \frac{\Delta}{2}\sigma^2 + \sigma\sqrt{\Delta}z_{ij}, \tag{9.12}$$

where z_{ij} is a random number generated from the standard normal distribution.

9.3   Generating Random Normal Numbers

The probability density function of a normal distribution is defined as (Evans et al., 2000, Chapter 30)

$$f(x, \mu, \sigma) = \frac{1}{\sqrt{2\pi}\sigma} \exp\left(-\frac{(x-\mu)^2}{2\sigma^2}\right), \quad x \in (-\infty, \infty), \tag{9.13}$$

where μ is the mean, σ is the standard deviation, and $\exp(\cdot)$ is the exponential function. The cumulative density function of the normal distribution is defined as

$$\begin{aligned} F(y, \mu, \sigma) &= \int_{-\infty}^{y} f(x, \mu, \sigma) dx \\ &= \int_{-\infty}^{y} \frac{1}{\sqrt{2\pi}\sigma} \exp\left(-\frac{(x-\mu)^2}{2\sigma^2}\right) dx, \quad y \in (-\infty, \infty). \end{aligned} \tag{9.14}$$

The cumulative density function of the standard normal distribution is $F(y, 0, 1)$.

There are several ways to generate random normal numbers. One way is to use the inversion method (Devroye, 1986). To use the inversion method to generate a standard normal number, we first generate a uniform random number u from $[0,1]$, and then we calculate the number z by solving the following equation:

$$u = F(z, 0, 1).$$

The number z is a standard random normal number.

9.4 Implementation

From Figure 9.1, we see that the input and the output are organized into the "Main" sheet and the "Scenarios" sheet, respectively. When the button is clicked, the scenarios are written to the "Scenarios" sheet. We organize the Visual Basic for Applications (VBA) code into two modules: the MGenerator module and the MInterface module.

9.4.1 MGenerator Module

This module contains VBA code for generating risk-neutral scenarios based on the Black–Scholes models described before. The declaration section of this module contains the following code:

```
1 Option Explicit
2
3 Private dSigma As Double
4 Private sHorizon As String
5 Private sTimeStep As String
6 Private lNumPath As Long
7 Private lSeed As Long
8 Private iPrecision As Integer
9 Private aFR() As Double
```

In the above code, we defined variables to store the parameters and the forward rates. These variables are private variables that can only be used by procedures in this module.

The following sub procedure, named Initialize, is used to initialize the variables declared above:

```
 1  Sub Initialize( _
 2      ByVal Sigma As Double , _
 3      ByVal Horizon As String , _
 4      ByVal TimeStep As String , _
 5      ByVal NumPath As Long , _
 6      ByVal Seed As Long , _
 7      ByVal Precision As Integer , _
 8      ByRef FR() As Double)
 9      dSigma = Sigma
10      sHorizon = Horizon
11      sTimeStep = TimeStep
12      lNumPath = NumPath
13      lSeed = Seed
14      iPrecision = Precision
15
16      Dim lL As Long
17      Dim lU As Long
18      lL = LBound(FR)
19      lU = UBound(FR)
20
21      ReDim aFR(1 To lU - lL + 1)
22      Dim i As Long
23      For i = lL To lU
24          aFR(i - lL + 1) = FR(i)
25      Next i
26
27  End Sub
```

In the above sub procedure, the forward rates are passed by reference. Other parameters are passed to the procedure by value. In Line 21, we re-dimension the module-level array aFR to hold the forward rates. In Lines 23–25, we copy the forward rates to the array.

To make sure the sub procedure Initialize works as expected, we need to test it. Before testing it, we need a sub procedure to display the contents of the module-level variables. We can write such a sub procedure as follows:

```
 1  Sub PrintInfo()
 2      Debug.Print "Volatility: " & CStr(dSigma)
 3      Debug.Print "Horizong: " & sHorizon
 4      Debug.Print "TimeStep: " & sTimeStep
 5      Debug.Print "NumPath: " & CStr(lNumPath)
 6      Debug.Print "Seed: " & CStr(lSeed)
```

```
 7      Debug.Print "Precision: " & CStr(iPrecision)
 8      Debug.Print "Number of Forward Rates: " & CStr(
            UBound(aFR) - LBound(aFR) + 1)
 9      Dim i As Long
10      Dim iCount As Integer
11      For i = LBound(aFR) To UBound(aFR)
12          Debug.Print Round(aFR(i), 6);
13          iCount = iCount + 1
14          If iCount >= 6 Then
15              Exit For
16          End If
17      Next i
18
19  End Sub
```

The sub procedure `PrintInfo` displays the values of the module-level variables in the Immediate window. Instead of displaying all the forward rates in the array, we only show the number of forward rates and the first six forward rates.

Now, we can use the following sub procedure to test the sub procedure `Initialize`:

```
 1  Sub TestInitialize()
 2      Dim rngParam As Range
 3      Dim rngFirstFR As Range
 4
 5      Set rngParam = Sheet1.Range("Param")
 6      Set rngFirstFR = Sheet1.Range("FirstFR")
 7
 8      Dim aFR() As Double
 9      Dim lFirstRow As Long
10      Dim lLastRow As Long
11
12      lFirstRow = rngFirstFR.Cells(1, 1).Row
13      lLastRow = rngFirstFR.Cells(1, 1).End(xlDown).Row
14
15      ReDim aFR(1 To lLastRow - lFirstRow + 1)
16      Dim i As Long
17      For i = 1 To lLastRow - lFirstRow + 1
18          aFR(i) = rngFirstFR.Cells(1, 1).Offset(i - 1,
                  0).Value
19      Next i
20
21      Initialize Sigma:=rngParam.Cells(1, 1).Value, _
```

```
22            Horizon:=rngParam.Cells(2, 1).Value, _
23            TimeStep:=rngParam.Cells(3, 1).Value, _
24            NumPath:=rngParam.Cells(4, 1).Value, _
25            Seed:=rngParam.Cells(5, 1).Value, _
26            Precision:=rngParam.Cells(6, 1).Value, _
27            FR:=aFR
28
29        PrintInfo
30 End Sub
```

The test sub procedure requires two named ranges, which are shown in Table 9.1. Since the number of forward rates varies, we only refer to the first forward rate. We assume that the forward rates are organized continuously in a column. In Lines 12–13, we calculate the first row and the last row of the forward rates. In Lines 17–19, we copy the forward rates from the worksheet to a procedure-level array named aFR. In Lines 21–27, we call the sub procedure Initialize with appropriate arguments. Line 29 calls the sub procedure PrintInfo to display the contents of the module-level variables. Executing the above code, we see the following output in the Immediate window:

```
1 Horizong: 30Y
2 TimeStep: 1M
3 NumPath: 100
4 Seed: 2016
5 Precision: 6
6 Number of Forward Rates: 360
7  0.00689   0.00689   0.00689   0.00689   0.00689   0.00689
```

Comparing the above output and Figure 9.1, we see that the sub procedure Initialize works well.

Table 9.1: Range Names Defined in the Interface of the Risk-Neutral Scenario Generator

Name	Refers To
FirstFR	=Main!$A$2
Param	=Main!$D$2:$D$7

Note that the forward rates depend on the time step. In the above test, we used a monthly time step. The forward rates are monthly forward rates obtained from the yield curve bootstrapper (see Chapter 8). If the time step is "1Y," we need to input annual forward rates, which can also be obtained from the yield curve bootstrapper by specifying the "Output Type" to "1Y."

To generate scenarios, we need first to figure out the number of time steps and the Δ required by Equation (9.10). To do that, we need a procedure to extract the number of periods and the period string from input strings, such as "30Y" and "1M." For this purpose, we can use the following sub procedure:

```vb
Sub ExtractPeriod(ByVal sIn As String, ByRef iNum As
    Integer, ByRef sPeriod As String)
    Dim iLen As Integer
    Dim sNum As String
    iLen = Len(sIn)

    If iLen < 2 Then
        Err.Raise Number:=1003, Source:="MHoliday.
            ExtractPeriod", Description:="Invalid
            period string"
    End If
    sPeriod = Right(sIn, 1)
    sNum = Left(sIn, iLen - 1)
    If IsNumeric(sNum) Then
        iNum = CInt(sNum)
    Else
        Err.Raise Number:=1003, Source:="MHoliday.
            ExtractPeriod", Description:="Invalid
            period string"
    End If
End Sub
```

The above sub procedure was first implemented in the yield curve bootstrapper. For an explanation of this sub procedure, readers are referred to Chapter 8.

Now, we can calculate the number of time steps using the following function:

```vb
Private Function CalNumTimeStep(ByVal Horizon As
    String, ByVal TimeStep As String) As Long
    Dim lRet As Long

    Dim iHorizonNum As Integer
    Dim sHorizonPeriod As String
    Dim iStepNum As Integer
    Dim sStepPeriod As String
```

```
8
9      ExtractPeriod sIn:=Horizon, iNum:=iHorizonNum,
           sPeriod:=sHorizonPeriod
10     ExtractPeriod sIn:=TimeStep, iNum:=iStepNum,
           sPeriod:=sStepPeriod
11
12     If Not StrComp(sHorizonPeriod, "Y", vbTextCompare
           ) = 0 Then
13         Err.Raise Number:=1001, Source:="MGenerator.
               CalNumTimeStep", _
14             Description:="Horizon should be specified
                   in years."
15     End If
16
17     If StrComp(sStepPeriod, "Y", vbTextCompare) = 0
           Then
18         lRet = iHorizonNum / iStepNum
19     ElseIf StrComp(sStepPeriod, "M", vbTextCompare) =
           0 Then
20         lRet = 12 * iHorizonNum / iStepNum
21     Else
22         Err.Raise Number:=1002, Source:="MGenerator.
               CalNumTimeStep", _
23             Description:="Unknown combination of
                   period strings."
24     End If
25
26     CalNumTimeStep = lRet
27 End Function
```

The function CalNumTimeStep is a private function and requires two arguments: a horizon string and a time step string. In the current version, the horizon can only be specified in years. The time step can be specified in months or years. The function will raise an error in other cases.

We can test the function CalNumTimeStep as follows:

```
1 Sub TestCalNumTimeStep()
2     Debug.Print CalNumTimeStep("30Y", "1Y")
3     Debug.Print CalNumTimeStep("30Y", "1M")
4     Debug.Print CalNumTimeStep("10Y", "3M")
5     Debug.Print CalNumTimeStep("30Y", "7M")
6 End Sub
```

Executing the above sub procedure gives

```
1  30
2  360
3  40
4  51
```

From the output, we see that the function calculates the number of time steps correctly.

To calculate the time step Δ, we use the following function:

```
1  Private Function CalDelta(ByVal TimeStep As String)
       As Double
2      Dim dRet As Double
3
4      Dim iStepNum As Integer
5      Dim sStepPeriod As String
6
7      ExtractPeriod sIn:=TimeStep, iNum:=iStepNum,
          sPeriod:=sStepPeriod
8
9      If StrComp(sStepPeriod, "Y", vbTextCompare) = 0
          Then
10         dRet = iStepNum
11     ElseIf StrComp(sStepPeriod, "M", vbTextCompare) =
          0 Then
12         dRet = iStepNum / 12#
13     Else
14         Err.Raise Number:=1004, Source:="MGenerator.
              CalDelta", _
15             Description:="Unknown time step string."
16     End If
17
18     CalDelta = dRet
19 End Function
```

The function CalDelta is also a private function that can only be used within this module. This version can only handle time steps specified in months or years. In Line 12, the sign # after 12 indicates that 12 is treated as a double-type number rather than an integer.

To test the function CalDelta, we use the following sub procedure:

```
1  Sub TestCalDelta()
2      Debug.Print CalDelta("1Y")
3      Debug.Print CalDelta("1M")
4      Debug.Print CalDelta("3M")
5  End Sub
```

Executing the above test sub procedure gives the following output:

```
1   1
2   8.33333333333333E-02
3   0.25
```

The output indicates that the function works well.

Now, we can implement the function for generating risk-neutral scenarios. We implement this function as follows:

```
1  Function Generate() As Variant
2      Dim lNumTimeStep As Long
3      Dim dDelta As Double
4
5      lNumTimeStep = CalNumTimeStep(sHorizon, sTimeStep
       )
6
7      If UBound(aFR) < lNumTimeStep Then
8          Err.Raise Number:=1005, Source:="MGenerator.
           Generate", _
9              Description:="There are not enough
               forward rates."
10     End If
11
12     dDelta = CalDelta(sTimeStep)
13
14     Dim aSce() As Double
15     ReDim aSce(1 To lNumPath, 1 To lNumTimeStep)
16
17     Rnd -1
18     Randomize lSeed
19
20     Dim i As Long
21     Dim j As Long
22     Dim z As Double
23     Dim dTmp As Double
24     For i = 1 To lNumPath
25         For j = 1 To lNumTimeStep
26             z = WorksheetFunction.NormInv(Rnd, 0, 1)
27             dTmp = WorksheetFunction.Ln(1 + dDelta *
               aFR(j)) - 0.5 * dDelta * dSigma *
               dSigma + _
28                 dSigma * Sqr(dDelta) * z
```

```
29              aSce(i, j) = Round(Exp(dTmp), iPrecision)
30          Next j
31      Next i
32
33      Generate = aSce
34 End Function
```

The function `Generate` creates a matrix of scenarios and returns it to the caller as a variant. In Lines 7–10, the function checks that there are enough forward rates. In Line 15, the function redimension a two-dimensional array to store the matrix of scenarios. In Lines 24–31, the function generates the scenarios according to the model given in Equation (9.10). We do not need to write a sub procedure to test this function, because we can test it using the button, which will be implemented in the module `MInterface`.

9.4.2 MInterface Module

The `MInterface` module contains VBA code for passing the parameters to the module `MGenerator` and writing the resulting scenarios to the "Scenarios" sheet.

In this module, we only need to implement a sub procedure for the button. We can implement the sub procedure as follows:

```
 1 Sub Button1_Click()
 2     Dim rngParam As Range
 3     Dim rngFirstFR As Range
 4
 5     Set rngParam = Sheet1.Range("Param")
 6     Set rngFirstFR = Sheet1.Range("FirstFR")
 7
 8     Dim aFR() As Double
 9     Dim lFirstRow As Long
10     Dim lLastRow As Long
11
12     lFirstRow = rngFirstFR.Cells(1, 1).Row
13     lLastRow = rngFirstFR.Cells(1, 1).End(xlDown).Row
14
15     ReDim aFR(1 To lLastRow - lFirstRow + 1)
16     Dim i As Long
17     For i = 1 To lLastRow - lFirstRow + 1
18         aFR(i) = rngFirstFR.Cells(1, 1).Offset(i - 1,
                0).Value
19     Next i
20
```

```
21    MGenerator.Initialize Sigma:=rngParam.Cells(1, 1)
          .Value, _
22          Horizon:=rngParam.Cells(2, 1).Value, _
23          TimeStep:=rngParam.Cells(3, 1).Value, _
24          NumPath:=rngParam.Cells(4, 1).Value, _
25          Seed:=rngParam.Cells(5, 1).Value, _
26          Precision:=rngParam.Cells(6, 1).Value, _
27          FR:=aFR
28
29    Dim vSce As Variant
30    vSce = MGenerator.Generate
31
32    ' write scenarios to Sheet2
33    Sheet2.Cells.ClearContents
34    Dim j As Long
35    For i = 1 To UBound(vSce, 1)
36        For j = 1 To UBound(vSce, 2)
37            Sheet2.Cells(i, j).Value = vSce(i, j)
38        Next j
39    Next i
40
41 End Sub
```

The code in the first part of this sub procedure is similar to that of the sub procedure TestInitialize implemented in the module MGenerator. In Lines 21–30, the sub procedure calls Initialize and Generate to generate scenarios. In Lines 35–39, it writes the scenarios to Sheet2.

We assign the sub procedure Button1_Click to the button in the interface. If we use the parameters and forward rates shown in Figure 9.1, we will get the scenarios shown in Figure 9.2.

Exercise 9.1. The volatility is a positive real number. However, we did not check the positivity of the volatility parameter in our implementation. Add code to the sub procedure Initialize to check the positivity of the volatility.

Exercise 9.2. The scenarios are saved to the worksheet as accumulation factors A_{ij} for $i = 1, 2, \ldots, n$ and $j = 1, 2, \ldots, m$, where n is the number of paths, and m is the number of time steps. Add code to the sub procedure Button1_Click to do the following:

- For each $j = 1, 2, \ldots, m$, calculate the mean μ_j and the standard deviation σ_j of $\ln A_{1j}, \ln A_{2j}, \ldots, \ln A_{nj}$, that is, calculate the mean and the

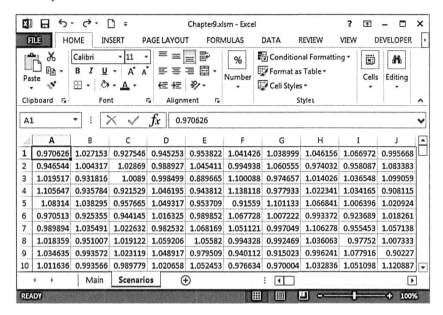

Figure 9.2: First few rows and columns of the risk-neutral scenarios

standard deviation of the continuous returns for each column of the scenarios.

- Write the annualized means $\frac{\mu_1}{\Delta}, \frac{\mu_2}{\Delta}, \dots, \frac{\mu_m}{\Delta}$ to the row that is two rows below the last row of the scenarios, and write the annualized volatilities $\frac{\sigma_1}{\sqrt{\Delta}}, \frac{\sigma_2}{\sqrt{\Delta}}, \dots, \frac{\sigma_m}{\sqrt{\Delta}}$ to the row that is three rows below the last row of the scenarios.

Exercise 9.3. Generating a large number of paths (e.g., $n = 5000$) takes some time. Add code to the function Generate in the module MGenerator to display the following message, "Generating path i of n" in the status bar, where i is the index of the path and n is the total number of paths. (Hint: use Application.StatusBar and DoEvents.)

9.5 Summary

In this chapter, we implemented a simple risk-neutral scenario generator based on the Black–Scholes model. The scenario generator generates risk-neutral scenarios for a single equity security. To generate scenarios for multiple equity securities with a correlation structure, readers are referred to (Carmona and Durrleman, 2005) and (Gan, 2015b). For more information about risk-neutral scenarios, readers are referred to (Sheldon and Smith, 2004).

10

Valuing a GMDB

In this chapter, we implement a simple program to calculate the fair market value and Greeks of a guaranteed minimum death benefit (GMDB) embedded in a variable annuity contract. The interface of the program is shown in Figure 10.1.

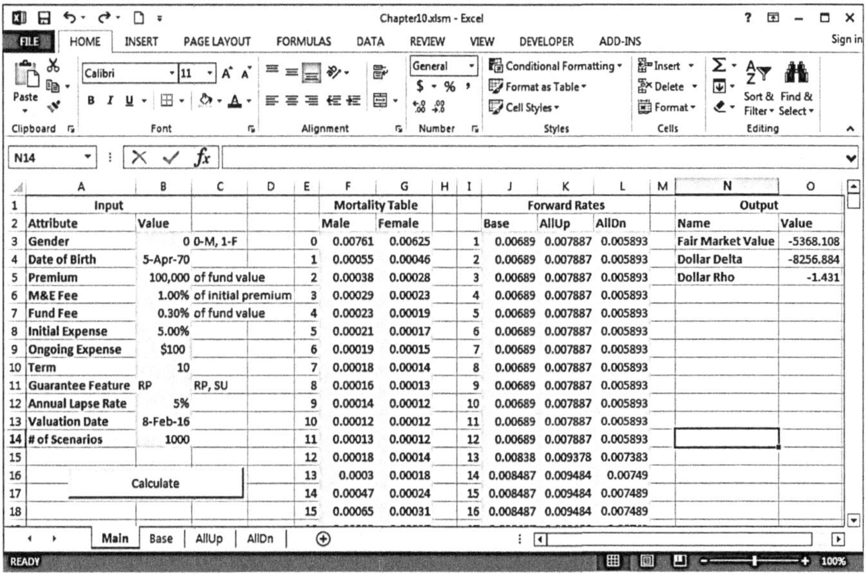

Figure 10.1: Interface of the GMDB valuation program

The inputs to the program include contract information, the mortality table, forward rates, and risk-neutral scenarios. The outputs include the fair market value, the dollar delta, and the dollar rho. The forward rates are obtained from the yield curve bootstrapper developed in Chapter 8. The risk-neutral scenarios are generated by the scenario generator developed in Chapter 9. We generate three sets of risk-neutral scenarios based on the base yield curve, the yield curve shifted 10bps [1] up, and the yield curve

---

[1] Here, "bps" means "basis points." A basis point is equal to 0.0001 or 0.01%.

shifted 10bps down, respectively. The three sets of scenarios are saved in the worksheets Base, AllUp, and AllDn, respectively.

10.1 Introduction

A variable annuity is a long-term life insurance contract between a policyholder and an insurance company under which the policyholder agrees to make a lump-sum purchase payment or a series of purchase payments to the insurance company, and, in return, the insurance company agrees to make benefit payments to the policyholder, beginning immediately or at a future date (The Geneva Association Report, 2013). Variable annuities are known as unit-linked products in Europe and segregated funds in Canada.

Under a variable annuity contract, the policyholder can allocate the account value to several subaccounts of the insurance company's separate account. The separate account of an insurance company is managed separately from the insurance company's general account and is protected from claims against the insurance company in the event of insolvency. Subaccounts are similar to mutual funds and allow the policyholder to participate in the stock market.

One main feature of variable annuities is that they contain guarantees, which include GMDB and guaranteed minimum living benefits (GMLB). The death benefit is paid to the beneficiary of a policyholder upon the death of the policyholder. There are several types of GMLB: guaranteed minimum accumulation benefits (GMAB), guaranteed minimum maturity benefits (GMMB), guaranteed minimum income benefits (GMIB), and guaranteed minimum withdrawal benefits (GMWB). The GMAB rider provides a guaranteed minimum survival benefit during a specified window after a specified waiting period, which is usually 10 years. The specified widows typically begins on an anniversary date and remains open for 30 days (Brown et al., 2002). The GMMB rider is similar to the GMAB rider, but the guaranteed amount is only available at the maturity of the contract. The GMIB rider gives a policyholder annuitized payments during a specified window after a specified waiting period. The annuitized payments are equal to the greater of payments determined by applying current fixed-annuitization rates to the payments determined by applying guaranteed fixed-annuitization rates, which are set at issue. The GMWB rider gives a policyholder the right to withdrawal a specified amount during the life of the contract until the initial investment is recovered.

Such guarantees are financial guarantees and cannot be addressed adequately by traditional actuarial approaches (Boyle and Hardy, 1997). Dynamic hedging is a popular approach to managing the financial risk associated with the guarantees and is adopted by some insurance companies. However, dynamic hedging requires calculating the Greeks (i.e., sensitivities of the value of guarantees on market factors) of the guarantees. Due to the complexity of guarantees, the Monte Carlo simulation method is usually used to value the guarantees. In the Monte Carlo simulation method, the cash flows of the guarantees are projected along many risk-neutral paths (see Chapter 9), and the value of the guarantees is calculated as the average present value of the cash flows.

The cash flows of different types of guarantees are projected differently, depending on the contract specifications. In this chapter, we introduce how to value a GMDB using the Monte Carlo simulation method.

10.2 Life Table Construction

Valuing a GMDB embedded in a variable annuity requires calculating mortalities. We use the US life tables 1999–2001 downloaded from `http://mort.soa.org/`. In these life tables, the mortality rates q_x are given for integer ages x from 0 to 109. Since valuing a GMDB involves fraction ages, we need to interpolate the mortalities of fraction ages from integer ages.

We use a linear method to interpolate the mortalities (Gupta and Varga, 2002). Let ℓ_n be the number of persons surviving to age n, where $n = 0, 1, \ldots, 109$. Then,

$$\ell_n = \prod_{i=0}^{n} (1 - q_i). \tag{10.1}$$

Let x be an arbitrary age. Then, we interpolate ℓ_x as follows:

$$\ell_x = (1 - \{x\})\ell_{[x]} + \{x\}\ell_{[x]+1}, \tag{10.2}$$

where $[x]$ and $\{x\}$ represent the integer part and the fractional part of x, respectively.

Under the linear interpolation method, $_t p_x$, the probability that a person aged x will survive to the age of $x + t$, is calculated as

$$_tp_x = \frac{\ell_{x+t}}{\ell_x} = \frac{(1 - \{x + t\})\ell_{[x+t]} + \{x + t\}\ell_{[x+t]+1}}{(1 - \{x\})\ell_{[x]} + \{x\}\ell_{[x]+1}}. \tag{10.3}$$

The probability that a person aged x will die before the age of $x + t$ is calculated as

$$_tq_x = 1 - _tp_x. \tag{10.4}$$

10.3   GMDB Valuation

Let F_{ij} $(i = 1, 2, \ldots, n, j = 1, 2, \ldots, m)$ denote the risk-neutral scenarios for the underlying fund that are generated by the risk-neutral scenario generator from Chapter 9. Here, n and m denote the number of paths and the number of time steps, respectively. In general, the larger the n and the m, the more accurate the results. However, the runtime increases when n and m increase. When using the Monte Carlo simulation method, there is a tradeoff between accuracy and runtime. In our example, we use $n = 1,000$ and $m = 360$ months.

For a variable annuity contract with a GMDB, we use the following notation to denote the cash flows of the GMDB that occur within the period $(t_{j-1}, t_j]$ along the ith risk-neutral path:

DB_{ij} denotes the claim of death benefit when the policyholder dies within the period $(t_{j-1}, t_j]$.

RC_{ij} denotes the risk charge for providing the guarantees.

In addition to the above notation, we use the following notation to denote the real and paper account values:

GD_{ij} denotes the guaranteed death benefit.

A_{ij} denotes the account value.

We use the following notation to denote various fees:

ϕ_{ME} denotes the annualized mortality and expense (M&E) fee of the contract. This fee is charged as a percentage of the fund value at the beginning of each anniversary.

ϕ_{IE} denotes the initial expense used to cover the sales commission. This fee is charged once at the inception of the policy.

ϕ_{OE} denotes the ongoing expense that is used to cover the administrative cost. This expense is a fixed dollar amount charged at the beginning of each anniversary.

ϕ_F denotes the annualized fund management fee of the underlying fund. Usually this fee is charged at the end of each period and goes to the fund manager rather than the insurance company.

The M&E fee includes the guarantee fee of the GMDB.

We can project the cash flows in a way that is similar to the way used by Bauer et al. (2008). For the sake of simplicity, we assume there is one type of event that will occur during the term of the contract: the policyholder passes away. We consider two GMDB riders:

DBRP contains a return-of-premium GMDB rider.

DBSU contains a step-up GMDB rider. The death benefit is reset to the account value annually if the latter is higher.

Suppose that the fees are charged from the account values at the end of every period. For $j = 0, 1, \ldots, m-1$, the cash flows from t_j to t_{j+1} are projected as follows:

- The account value evolves as follows:

$$A_{i,j+1} = \begin{cases} A_{ij}F_{i,j+1}\left(1-\Delta\phi_F\right)\left(1-\phi_{ME}\right) - \phi_{OE}, & \text{if } t_{j+1} \text{ is an anniversary} \\ A_{ij}F_{i,j+1}\left(1-\Delta\phi_F\right), & \text{if otherwise,} \end{cases}$$

(10.5)

where $\Delta = t_{j+1} - t_j$. Here, we assume that the fund management fees are deducted at the end of each period. Here, we consider only single-premium products. The fund value at inception is calculated as follows:

$$A_{i,0} = \text{Premium} \times (1 - \phi_{IE})(1 - \phi_{ME}) - \phi_{OE}.$$

- The risk charges are projected as

$$RC_{i,j+1} = \begin{cases} A_{ij}F_{i,j+1}\left(1-\Delta\phi_F\right)\phi_{ME}, & \text{if } t_{j+1} \text{ is an anniversary} \\ 0, & \text{if otherwise.} \end{cases}$$

(10.6)

The risk charge at inception is calculated as

$$RC_{i,0} = \text{Premium} \times (1 - \phi_{IE})\phi_{ME}.$$

- If the contract contains a return-of-premium GMDB rider, the death benefit is equal to the initial premium:

$$GD_{i,j+1} = GD_{i,j}, \quad GD_{i,0} = \text{Premium}. \tag{10.7}$$

- If the contract contains a step-up GMDB rider, we have

$$GD_{i,j+1} = \begin{cases} \max\{GD_{i,j}, A_{i,j+1}\}, & \text{if } t_{j+1} \text{ is an anniversary.} \\ GD_{i,j}, & \text{if } t_{j+1} \text{ is not an anniversary,} \end{cases} \tag{10.8}$$

with $GD_{i,0} = \text{Premium}$.

- If the policyholder dies within the period $(t_j, t_{j+1}]$, then the claim of death benefit is projected as

$$DB_{i,j+1} = \max\{0, GD_{i,j+1} - A_{i,j+1}\}. \tag{10.9}$$

- After the maturity of the contract, all the state variables are set to zero.

Once we have all the cash flows, we can calculate the fair market values of the riders as follows:

$$V_0 = \frac{1}{n} \sum_{i=1}^{n} \sum_{j=1}^{m} {}_{(j-1)\Delta}p_{x_0} \cdot {}_{\Delta}q_{x_0+(j-1)\Delta} L_j DB_{i,j} d_j, \tag{10.10}$$

where x_0 is the age of the policyholder, p is the survival probability, q is the probability of death, r_L is the annual lapse rate, L_j is the lapse rate calculated as

$$L_j = \begin{cases} (1 - \Delta r_L)^j, & \text{if } t_j < T, \\ 0, & \text{if } t_j \geq T, \end{cases}$$

and d_j is the discount factor defined as

$$d_j = \exp\left(-\Delta \sum_{l=1}^{j} f_l\right).$$

Here, r_L is the annual lapse rate, T is the maturity of the policy, and $f_1, f_2, \ldots, f_m$ are forward rates. The risk charge value is calculated as

$$RC_0 = \frac{1}{n} \sum_{i=1}^{n} \sum_{j=1}^{m} {}_{j\Delta}p_{x_0} RC_{i,j} L_j d_j. \tag{10.11}$$

The fair market value is then calculated as

$$FMV_0 = V_0 - RC_0. \tag{10.12}$$

10.4 Greek Calculation

Hedging the financial risks associated with the guarantees usually requires calculating the Greeks of the guarantees. The Greeks are the sensitivities of the fair market value of the guarantees to market variables, such as equity and the interest rate. The sensitivity to equity is referred to as the delta, while the sensitivity to the interest rate is referred to as rho.

There are several ways to calculate the Greeks (Cathcart et al., 2015; Fu, 2015). Here, we use the finite difference method to calculate the Greeks. To calculate the delta of the GMDB, we shock the initial account values $A_{i,0}$ up 1% and calculate the fair market value FMV_u of the GMDB. Then we shock the initial account values $A_{i,0}$ down 1% and calculate the fair market value FMV_d of the GMDB. Then, we calculate the dollar delta as

$$\$Delta = \frac{FMV_u - FMV_d}{0.02}.$$

The dollar delta approximates the dollar value change of the fair market value of the guarantee when the equity increases 1%.

Calculating the rho is a little bit tedious. First, we shock the swap rates up 10bps (i.e., 0.001) and bootstrap the shocked swap rates to get the forward rates, which are used to generate risk-neutral scenarios. Then, we calculate the fair market value FMV_u of the guarantee based on the risk-neutral scenarios. Second, we shock the swap rates down 10bps (i.e., −0.001) and bootstrap the shocked swap rates to get the forward rates. We generate risk-neutral scenarios and calculate the fair market value FMV_d of the guarantee again based on the new scenarios. The dollar rho is calculated as

$$\$Rho = \frac{FMV_u - FMV_d}{20}.$$

The dollar rho approximates the dollar value change of the fair market value of the guarantee when the yield curve shifts up 1bps.

10.5 Implementation

We organize the Visual Basic for Applications (VBA) code of the valuation program into three modules: the MLifeTable module, the MGMDB module, and the MInterface module. A list of range names used by the program is given in Table 10.1.

Table 10.1: Range Names Used by the GMDB Valuation Program

Name	Refers To
FRAllDn	=Main!$L$3:$L$362
FRAllUp	=Main!$K$3:$K$362
FRBase	=Main!$J$3:$J$362
MTFemale	=Main!$G$3:$G$112
MTMale	=Main!$F$3:$F$112
Param	=Main!$B$3:$B$13

10.5.1 MLifeTable Moduel

The `MLifeTable` module contains code for interpolating the life table. The declarations section of this module contains the following code:

```
 1  Option Explicit
 2
 3  Private aLT() As Double
 4  Private iMaxAge As Integer
```

In this module, we declared two private module-level variables. The first variable is used to store the mortality rates shown in Figure 10.1. The second variable is used to store the maximum age available in the mortality table.

We need a procedure to create the life table from the mortality rates. We implement the procedure as follows:

```
 1  Sub Initialize(ByRef rngMTMale As Range, ByRef
        rngMTFemale As Range)
 2      Dim nRow As Integer
 3
 4      nRow = rngMTMale.Rows.Count
 5      If nRow <> rngMTFemale.Rows.Count Then
 6          Err.Raise Number:=1001, Source:="MLifeTable.
                Initialize", Description:="Input ranges
                have different size."
 7      End If
 8
 9      iMaxAge = nRow - 1
10      ReDim aLT(0 To nRow - 1, 0 To 1)
11
12      ' convert mortality rates to survival
            probabilities
13      Dim i As Integer
```

```
14      aLT(0, 0) = 1 - rngMTMale.Cells(i + 1, 1).Value
15      aLT(0, 1) = 1 - rngMTFemale.Cells(i + 1, 1).Value
16      For i = 1 To nRow - 1
17          aLT(i, 0) = aLT(i - 1, 0) * (1 - rngMTMale.
                Cells(i + 1, 1).Value)
18          aLT(i, 1) = aLT(i - 1, 1) * (1 - rngMTFemale.
                Cells(i + 1, 1).Value)
19      Next i
20 End Sub
```

The sub procedure requires two arguments: a range containing male mortality rates, and a range containing female mortality rates. In Lines 4–7, we check whether the two input ranges have the same number of rows. If the two ranges do not have the same number of rows, an error will be raised. The maximum available age in the mortality rates is the number of rows minus one, as the mortality rates start from age 0. In Line 10, we redimension the array aLT so that it can hold all the survival probabilities, which are calculated in Lines 14–19.

Exercise 10.1. Write a sub procedure named PrintInfo to display the values of the module-level variables in the Immediate window.

Exercise 10.2. Write a sub procedure named TestInitialize to test the sub procedure Initialize. In particular, you first call the sub procedure Initialize and then call PrintInfo (see Exercise 10.1) to show the values of the module-level variables.

We need to implement some functions to calculate the survival probability for fractional ages. We first implement the interpolation method described in Section 10.2 as follows:

```
1 Function Interpolate(ByVal x As Double, ByVal g As
      Integer) As Double
2      Dim iL As Integer
3      Dim iU As Integer
4      Dim i As Integer
5
6      If x < 0 Or x > iMaxAge Then
7          Err.Raise Number:=1002, Source:="MLifeTable.
              Interpolate", Description:="Age x is out
              of range."
```

```
 8        End If
 9
10        iL = WorksheetFunction.Floor(x, 1)
11        iU = iL + 1
12
13        Interpolate = (iU - x) * aLT(iL, g) + (x - iL) *
              aLT(iU, g)
14 End Function
```

In Lines 6–8 of the above code, we first check that the input age is not less than zero or larger than the maximum age. Since we use the linear interpolation method, we cannot extrapolate the survival probabilities. In Line 10, we get the integer part of the input age. The code in Line 13 implements the linear interpolation method based on Equation (10.2).

To test the function `Interpolate`, we use the following sub procedure:

```
 1 Sub TestInterpolate()
 2        Dim rngMTMale As Range
 3        Dim rngMTFemale As Range
 4
 5        Set rngMTMale = Range("MTMale")
 6        Set rngMTFemale = Range("MTFemale")
 7
 8        Initialize rngMTMale:=rngMTMale, rngMTFemale:=
              rngMTFemale
 9
10        Debug.Print Interpolate(0, 0)
11        Debug.Print Interpolate(35.5, 0)
12        Debug.Print Interpolate(35.5, 1)
13        Debug.Print Interpolate(99.1, 0)
14 End Sub
```

In the above test procedure, we used two named ranges defined in Table 10.1. We calculated the survival probabilities for a male aged 0, 35.5, and 99.1. We also calculated the survival probability for a female aged 35.5. Executing the above sub procedure gives the following output:

```
1   0.99239
2   0.961165379370226
3   0.979048981352752
4   6.21200647276638E-03
```

From the output, we see that a female aged 35.5 has a higher survival probability than a male of the same age.

Now, we can implement the functions for calculating $_t p_x$ and $_t q_x$ for arbitrary t and x. We implement the two functions as follows:

```
1 Function tpx(ByVal x As Double, ByVal t As Double,
      ByVal g As Integer) As Double
2     tpx = Interpolate(x + t, g) / Interpolate(x, g)
3 End Function
4
5 Function tqx(ByVal x As Double, ByVal t As Double,
      ByVal g As Integer) As Double
6     tqx = 1 - tpx(x, t, g)
7 End Function
```

In the above code, we implemented Equations (10.3) and (10.4) straightforwardly.

———————————————◦———————————————

Exercise 10.3. Write a sub procedure named Testtpx to test the function tpx by calculating the following probabilities: $_1 p_{35}$ and $_{3.6} p_{35.2}$ for a male, and $_{3.6} p_{35.2}$ for a female.

———————————————◦———————————————

10.5.2 MGMDB Moduel

The MGMDB module implements the GMDB valuation model described in Section 10.3. The declarations section of this module contains the following code:

```
1  Option Explicit
2
3  Private iGender As Integer
4  Private datDOB As Date
5  Private dPremium As Double
6  Private dMEFee As Double
7  Private dFundFee As Double
8  Private dIE As Double
9  Private dOE As Double
10 Private iTerm As Integer
11 Private sFeature As String
12 Private dLapseRate As Double
13 Private datValDate As Date
14 Private iNumSce As Integer
```

```
15 Private rngFR As Range
16 Private shtSce As Worksheet
17
18 Private Const dTimeStep As Double = 1# / 12
19 Private ap(0 To 360) As Double ' ap(i) = x survives
      to (i*dTimeStep)
20 Private aq(1 To 360) As Double ' aq(i) = x dies in ((
      i-1)*dTimeStep, i*dTimeStep)
21 Private ad(0 To 360) As Double
22 Private aL(0 To 360) As Double
23
24 Private aA(0 To 360) As Double
25 Private aRC(0 To 360) As Double
26 Private aGD(0 To 360) As Double
27 Private aDB(0 To 360) As Double
```

In Lines 3–14 of the above code, we declared some variables to hold the
parameters of the contract. In Lines 15–16, we declared two variables to hold
the references to the range of forward rates and the worksheet of scenarios.
In Line 18, we defined a constant for the time step. In this program, we only
consider monthly time steps. In Lines 19–22, we defined some arrays to
hold the survival probabilities, the probabilities of dying within a period,
the discount factors, and the survivorship. In Lines 24–27, we declared some
variables to hold temporary values. The names of the variables match those
used in the formulas described in Section 10.3.

Valuing a GMDB requires two sets of information: the contract and the
scenarios. Hence, we implement two procedures to initialize the module-
level variables. We first implement the contract initialization procedure as
follows:

```
1  Sub InitializeContract( _
2     ByVal Gender As Integer, _
3     ByVal DOB As Date, _
4     ByVal Premium As Double, _
5     ByVal MEFee As Double, _
6     ByVal FundFee As Double, _
7     ByVal IE As Double, _
8     ByVal OE As Double, _
9     ByVal Term As Integer, _
10    ByVal Feature As String, _
11    ByVal LapseRate As Double, _
12    ByVal ValDate As Date, _
13    ByVal NumSce As Integer)
14    iGender = Gender
```

```
15    datDOB = DOB
16    dPremium = Premium
17    dMEFee = MEFee
18    dFundFee = FundFee
19    dIE = IE
20    dOE = OE
21    iTerm = Term
22    sFeature = Feature
23    dLapseRate = LapseRate
24    datValDate = ValDate
25    iNumSce = NumSce
26
27    Dim dx As Double
28    dx = DateDiff("d", datDOB, datValDate) / 365#
29    Dim i As Integer
30    ap(0) = 1
31    aL(0) = 1
32    For i = 1 To 360
33        ap(i) = MLifeTable.tpx(dx, i * dTimeStep,
              iGender)
34        aq(i) = ap(i - 1) * MLifeTable.tqx(dx + (i -
              1) * dTimeStep, dTimeStep, iGender)
35        If i < iTerm * 12 Then
36            aL(i) = aL(i - 1) * (1 - dTimeStep *
                  dLapseRate)
37        Else
38            aL(i) = 0
39        End If
40    Next i
41 End Sub
```

The sub procedure `InitializeContract` requires quite a few arguments, which correspond to the parameters of the contract and the valuation model. In Lines 14–25, we initialize the variables using the input values. In Line 28, we calculate the age of the contract holder by dividing the number of days between the date of birth and the valuation date by 365. In Lines 32-40, we calculate the survival probabilities, the probability of dying within each period, and the survivorship.

The second initialization procedure is simple and is implemented as follows:

```
1 Sub InitializeScenario( _
2     ByRef Sce As Worksheet, _
3     ByRef FR As Range)
```

```
4     Set shtSce = Sce
5     Set rngFR = FR
6
7     Dim i As Integer
8     ad(0) = 1
9     For i = 1 To 360
10        ad(i) = ad(i - 1) * Math.Exp(-dTimeStep *
              rngFR.Cells(i, 1))
11    Next i
12 End Sub
```

This sub procedure requires only two arguments: the range of forward rates and the worksheet of scenarios. Since the arguments are objects, we pass the arguments by reference. In this sub procedure, we also calculate the discount factors of all periods that can be used later.

Once we have all the contract information and the scenarios, we can project the cash flows of the guarantee and calculate the present values. We first implement a function to project the cash flows along a single path. We implement this function as follows:

```
1  Function Project(ByVal dShock As Double, ByVal ind As
       Integer) As Double
2      Dim i As Integer
3
4      aA(0) = (1 + dShock) * (dPremium * (1 - dIE) * (1
           - dMEFee) - dOE)
5      aRC(0) = dPremium * (1 - dIE) * dMEFee
6      aGD(0) = dPremium
7
8      For i = 1 To 360
9          If i Mod 12 = 0 Then
10             aA(i) = aA(i - 1) * shtSce.Cells(ind, i).
                   Value * _
11                 (1 - dTimeStep * dFundFee) * (1 -
                       dMEFee) - dOE
12             aRC(i) = aA(i - 1) * shtSce.Cells(ind, i)
                   .Value * _
13                 (1 - dTimeStep * dFundFee) * dMEFee
14             If StrComp(sFeature, "RP", vbTextCompare)
                   = 0 Then
15                 aGD(i) = aGD(i - 1)
16             Else
17                 aGD(i) = WorksheetFunction.Max(aGD(i
                       - 1), aA(i))
```

```
18                     End If
19                 Else
20                     aA(i) = aA(i - 1) * shtSce.Cells(ind, i).
                          Value * _
21                         (1 - dTimeStep * dFundFee)
22                     aRC(i) = 0
23                     aGD(i) = aGD(i - 1)
24                 End If
25                 aDB(i) = WorksheetFunction.Max(0, aGD(i) - aA
                      (i))
26             Next i
27
28         Dim dPVBenefit As Double
29         Dim dPVCharge As Double
30         For i = 1 To 360
31             dPVBenefit = dPVBenefit + aq(i) * aL(i) * aDB
                  (i) * ad(i)
32             dPVCharge = dPVCharge + ap(i) * aL(i) * aRC(i
                  ) * ad(i)
33         Next i
34         Project = dPVBenefit - dPVCharge
35     End Function
```

The function Project requires two arguments: an equity shock amount and an index of scenario. The equity shock argument is used to calculate the dollar delta described in Section 10.4. In Lines 4–6, we calculate the account value, the risk charge, and the guaranteed benefit at time 0. In Lines 8–26, we project the cash flows for 360 months. In Lines 10–18, we project the cash flows when the time is an anniversary. Lines 20–23 project the cash flows when the time is not an anniversary. The death benefit for each month is calculated in Line 25. In Lines 30–34, we calculate the present value of the cash flows.

Exercise 10.4. Write a sub procedure named TestProject to test the function Project by doing the following:

- Add a new worksheet named Temp if it does not exist.
- Initialize the modules MLifeTable and MGMDB.
- Call the function Project with dShock=0 and ind=1.
- Write all the module-level arrays to the worksheet Temp, such that each

row contains an array, and the name of the array appears in the first column.

Now, we implement the function to calculate the fair market value of the GMDB as follows:

```
1  Function FMV(ByVal dShock As Double) As Double
2      Dim dSum As Double
3      Dim i As Integer
4
5      For i = 1 To iNumSce
6          dSum = dSum + Project(dShock, i)
7      Next i
8
9      FMV = dSum / iNumSce
10 End Function
```

In the above function, we calculate the average present value by calling the function Project for the specified number of scenarios.

10.5.3   MInterface Module

The MInterface module contains the VBA macro for calling various functions to calculate the fair market value, the dollar delta, and dollar rho of a GMDB embedded in a variable annuity contract. This module also contains code to write the results to the worksheet. We implement the module as follows:

```
1  Option Explicit
2
3  Sub Button1_Click()
4      Dim dFMVBase As Double
5      Dim dFMVEU As Double
6      Dim dFMVED As Double
7      Dim dFMVIU As Double
8      Dim DFMVID As Double
9
10     MLifeTable.Initialize rngMTMale:=Range("MTMale"), _
           rngMTFemale:=Range("MTFemale")
11     With Range("Param")
12         MGMDB.InitializeContract _
13             Gender:=.Cells(1, 1).Value, _
14             DOB:=.Cells(2, 1).Value, _
```

```
15              Premium:=.Cells(3, 1).Value, _
16              MEFee:=.Cells(4, 1).Value, _
17              FundFee:=.Cells(5, 1).Value, _
18              IE:=.Cells(6, 1).Value, _
19              OE:=.Cells(7, 1).Value, _
20              Term:=.Cells(8, 1).Value, _
21              Feature:=.Cells(9, 1).Value, _
22              LapseRate:=.Cells(10, 1).Value, _
23              ValDate:=.Cells(11, 1).Value, _
24              NumSce:=.Cells(12, 1).Value
25      End With
26
27      MGMDB.InitializeScenario Sce:=Sheets("Base"), FR
            :=Range("FRBase")
28      dFMVBase = MGMDB.FMV(0)
29      dFMVEU = MGMDB.FMV(0.01)
30      dFMVED = MGMDB.FMV(-0.01)
31
32      MGMDB.InitializeScenario Sce:=Sheets("AllUp"), FR
            :=Range("FRAllUp")
33      dFMVIU = MGMDB.FMV(0)
34
35      MGMDB.InitializeScenario Sce:=Sheets("AllDn"), FR
            :=Range("FRAllDn")
36      DFMVID = MGMDB.FMV(0)
37
38      Sheet1.Range("O3").Value = dFMVBase
39      Sheet1.Range("O4").Value = (dFMVEU - dFMVED) /
            0.02
40      Sheet1.Range("O5").Value = (dFMVIU - DFMVID) / 20
41 End Sub
```

In the above function, we first initialize the module MLifeTable in Line 10. Then, we initialize the contract information in the module MGMDB in Lines 11–25. In Line 27, we pass the worksheet containing the base scenarios and the range containing the corresponding forward rates to the module MGMDB. In Lines 28–30, we calculate the fair market values for different equity shocks. In Lines 32–36, we calculate the fair market values for different interest rate shocks. Finally we write the results to the worksheets in Lines 38–40.

If we use the inputs shown in Figure 10.1, we see that outputs shown in the same figure. The fair market value of this GMDB rider is –$5368.11. Since it is negative, the present value of the benefits is lower than that of the fees. The dollar delta is –$8256.88, meaning that when the market goes up 1%, the fair market value of the guarantee will decrease about –$82.5688. The dollar rho is close to zero, meaning that the fair market value of the GMDB guarantee is not sensitive to the yield curve changes.

Exercise 10.5. Recalculate the output by changing the guarantee feature from Return of Premium (RP) to Step-Up (SU). Which guarantee is more expensive between a GMDB with RP and a GMDB with SU, holding all other parameters constant?

Exercise 10.6. Modify the sub procedure `Button1_Click` by adding code to calculate the runtime of the valuation program. Write the runtime to the cell "O6" of the main worksheet. (Hint: Use the `Timer` function.)

10.6   Summary

Variable annuities are insurance products that contain guarantees. Since these guarantees are complicated, closed-form formulas to calculate the fair market values of these guarantees do not exist. Insurance companies rely heavily on Monte Carlo simulation to calculate the fair market values of guarantees. In this chapter, we implemented a simple Monte Carlo method to calculate the fair market value of a GMDB.

An advantage of the Monte Carlo simulation method is that it is flexible to value various guarantees. However, Monte Carlos simulation is time-consuming, especially when used to value a large portfolio of variable annuities. Metamodeling approaches have been proposed to speed up the Monte Carlo simulation method. See (Gan, 2013), (Gan, 2015a), (Gan and Lin, 2015), and (Gan and Lin, 2016) for details.

11

Connecting to Databases

In this chapter, we implement a simple tool in Visual Basic for Applications (VBA) to extract variable annuity contracts from an Access® database that are issued in a particular year. This tool also allows us to save extracted contracts into a new table in the database. The interface of this inforce tool is shown in Figure 11.1.

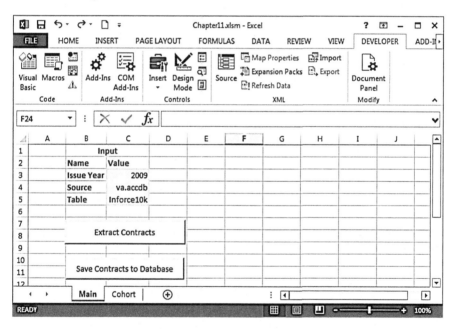

Figure 11.1: Interface of the variable annuity inforce tool

Table 11.1: Range Names Used by the Variable Annuity Inforce Tool

Name	Refers To
FileName	=Main!$C$4
IssueYear	=Main!$C$3
TableName	=Main!$C$5

Table 11.1 gives a list of range names used by the simple tool. Figure 11.2 shows the variable annuity inforce table in an Access® database named "va.accdb." The name of the table is "Inforce10k," and the table contains 10,000 variable annuity contracts, which are issued in different time. These variable annuity contracts are generated randomly by a Java program (Gan, 2015b).

recordID	survivorShip	gender	prodType	issueDate	matDate	birthDate
1	1	M	DBRP	3/24/2008	3/24/2037	5/12/1972
2	1	M	DBRU	12/15/2008	12/15/2037	2/9/1978
3	1	M	DBRU	10/9/2012	10/12/2032	7/6/1969
4	1	F	MB	11/27/2000	11/27/2026	12/4/1952
5	1	M	DBRU	4/30/2001	4/30/2024	7/17/1961
6	1	F	DBRU	9/13/2006	9/13/2034	2/10/1950
7	1	F	WBSU	5/28/2004	5/28/2030	11/22/1973
8	1	M	DBRP	4/3/2002	4/3/2020	6/24/1955
9	1	M	WBSU	3/29/2010	3/29/2033	10/31/1957
10	1	F	WB	4/26/2007	4/26/2035	12/23/1970
11	1	F	MB	4/8/2009	4/8/2037	10/7/1979
12	1	F	MB	11/14/2011	11/14/2039	3/26/1972

Figure 11.2: Inforce10k table in an Access® database

11.1 ActiveX® Data Objects

Excel® is a powerful tool for analyzing data and producing customized reports. However, Excel may fall short as a data storage application when dealing with extremely large volumes of data. Databases are usually used to store such large volumes of data.

A database is an organized collection of data and is created and managed by a database management system (DBMS), which is a computer software application that provides users and programmers with a systematic way to create, retrieve, update, and manage data. Popular database management systems include Microsoft Access, Microsoft SQL Server, Oracle®, and MySQL®.

Excel VBA provides some methods to access databases. One method to access databases in Excel VBA is through Microsoft's ActiveX Data Objects (ADO). ADO consists a set of objects for accessing data sources. A main advantage of ADO is that it allows application developers to write programs to access data without knowing how the database is implemented. In particular, ADO allows us to do the following:

- Connect to various external databases.

- Add, delete, and edit records from a database.

- Query data from a database to get a recordset.

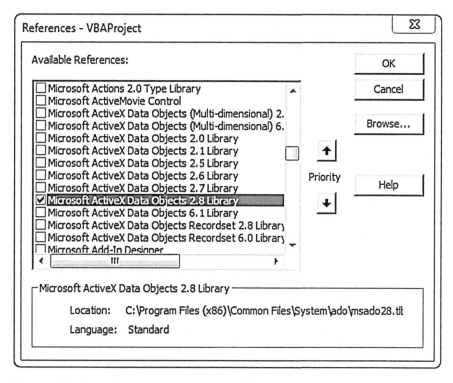

Figure 11.3: Checking the reference for Microsoft ActiveX Data Objects 2.8 Library

To use ADO to access databases in Excel VBA, we need to select "Microsoft ActiveX Data Objects 2.8 Library" in the References window (see Figure 11.3). To show the References window, you can click "Tools/References" in the Visual Basic Editor.

In this chapter, we will use the following objects of ADO: the Connection object, the RecordSet object, and the Command object. The Connection object is used to establish a connection between Excel and a database. The RecordSet object is used to retrieve a single record or multiple records from a database table. The RecordSet object can also be used to save records from Excel to a database table. The Command object is used to hold information related to queries to be executed.

11.2 SQL Queries

SQL stands for Structured Query Language, which is a database language used for querying, updating, and managing relational databases. SQL is a complex language. In this section, we introduce some commonly used SQL statements.

To retrieve data from one ore more tables, we use the SELECT statement. The SQL expression to retrieve specified columns and all rows from a table is as follows:

```
1  SELECT [ColumnName1], [ColumnName2] FROM TableName
```

The column names are separated by commas. If a column name contains spaces or is an SQL keyword, we need to enclose the column name with square brackets "[]."

If we want to retrieve all columns and all rows from a table, we use the following simplified expression:

```
1  SELECT * FROM TableName
```

We use * to replace all the column names.

To retrieve specified columns and specified rows from a table, we use the following SQL expression:

```
1  SELECT [ColumnName1], [ColumnName2] FROM TableName
      WHERE Condition
```

To retrieve specified rows, we add a WHERE clause to specify the condition for the rows. To sort the retrieved data, we can add an ORDER BY clause at the end:

```
1  SELECT [ColumnName1], [ColumnName2] FROM TableName
      WHERE Condition ORDER BY [ColumnName1]
```

To delete a table in a database, we use the following SQL expression:

```
1 DROP TABLE TableName
```

The SQL expression used to create a table in a database is given by

```
1 CREATE TABLE TableName (ColumnName1 DataType ,
      ColumnName2 DataType , ColumnName3 DataType);
```

There are many other SQL statements (Forta, 2012; Tale, 2016). The SQL statements mentioned above will be used in this chapter.

11.3   Implementation

We organize the VBA code of the inforce tool into two modules: the MDatabase module and the MInterface module.

11.3.1   MDatabase Module

The MDatabase module contains VBA code to access databases. The declaration section of this module contains the following code:

```
1 Option Explicit
2
3 Dim dbConnection As ADODB.Connection
```

In the declaration section, we define only one module-level variable, which is an ADODB.Connection object. This variable is used to hold the connection to a database.

The module-level variable is initialized by the following sub procedure:

```
1 Sub Connect(ByVal FileName As String)
2     Dim sConn As String
3     sConn = "Provider=Microsoft.ACE.OLEDB.12.0;Data
          Source=" & _
4         FileName & ";Persist Security Info=False;"
5
6     Set dbConnection = New ADODB.Connection
7     dbConnection.ConnectionString = sConn
8     dbConnection.Open
9 End Sub
```

This sub procedure requires one argument, which is the name of the Access database. In Line 3, we construct a connection string by specifying the Provider, the Data Source, and the Persist Security Info. The code in Line 3 is a common syntax for version 2007 and later versions of Excel. We set Persist Security Info to false so that security-sensitive information (e.g., the password) will not be returned as part of the connection. In Line 6, we create a new ADODB.Connection object and assign it to the module-level variable. In Line 7, we assign the connection string to the object. In Line 8, we open the connection.

To close the connection, we use the following sub procedure:

```
1  Sub  Disconnect ()
2      dbConnection.Close
3      Set  dbConnection  =  Nothing
4  End  Sub
```

In Line 2 of the above code, we call the Close method to close the connection. In Line 3, we release the memory of the variable by setting the variable to Nothing.

The following function is used to retrieve a list of tables stored in an Access database:

```
1  Function  ListTables ()
2      Dim  dbRecordSet  As  ADODB.Recordset
3
4      Set  dbRecordSet  =  dbConnection.OpenSchema (
           adSchemaTables )
5
6      Dim  vTmp  As  Variant
7      vTmp  =  dbRecordSet.GetRows
8
9      Dim  nRow  As  Integer
10     nRow  =  UBound(vTmp ,  2)
11
12     Dim  aTable()  As  String
13     ReDim  aTable(0  To  nRow)
14
15     Dim  i  As  Integer
16     Dim  j  As  Integer
17     For  i  =  0  To  UBound(vTmp ,  1)
18         If  StrComp(dbRecordSet.Fields(i).Name ,  "
                TABLE_NAME" ,  vbTextCompare)  =  0  Then
19             For  j  =  0  To  UBound(vTmp ,  2)
20                 aTable(j)  =  vTmp(i ,  j)
```

```
21              Next j
22              Exit For
23          End If
24      Next i
25
26      ListTables = aTable
27 End Function
```

In Line 2 of the above code, we declare an ADODB.Recordset object variable. In Line 4, we call the OpenSchema method to get schema of the database and assign it to the ADODB.Recordset object variable. The schema of the database contains the names of the tables stored in the database. In Line 7, we convert the record set into a two-dimensional array by calling the GetRows method. In Lines 17–24, we extract the table names and put them into a one-dimensional array.

To test the function ListTables, we use the following sub procedure:

```
1 Sub TestListTables()
2     Dim sFileName As String
3
4     sFileName = ThisWorkbook.Path & "\va.accdb"
5     Connect (sFileName)
6
7     Dim vRes As Variant
8     vRes = ListTables
9
10    Disconnect
11
12    Dim i As Integer
13    For i = 0 To UBound(vRes)
14        Debug.Print vRes(i)
15    Next i
16 End Sub
```

In the above test procedure, we assume that the Access database va.accdb is saved in the same folder as the workbook. In Line 5, we open the connection by calling the procedure Connect. In Line 8, we call the function ListTables to get the table names. In Line 10, we close the connection by calling the procedure Disconnect. Executing the above test procedure gives the following output:

```
1 Inforce10k
2 MSysAccessStorage
3 MSysACEs
```

```
 4 MSysComplexColumns
 5 MSysIMEXColumns
 6 MSysIMEXSpecs
 7 MSysNameMap
 8 MSysNavPaneGroupCategories
 9 MSysNavPaneGroups
10 MSysNavPaneGroupToObjects
```

From the output, we see that many names start with "MSys." These tables
are Microsoft Access system tables, which contain general information
about various specifications. The first name "Inforce10k" is the name of the
variable annuity inforce table.

Before creating a table in a database, we need to make sure that the
table does not exist in the database. The following function can be used to
detect whether a table exists or not in the database:

```
 1 Function HasTable(ByVal TableName As String)
 2     Dim vTables As Variant
 3     Dim bRes As Boolean
 4
 5     vTables = ListTables
 6     bRes = False
 7     Dim i As Integer
 8     For i = 0 To UBound(vTables)
 9         If StrComp(vTables(i), TableName,
             vbTextCompare) = 0 Then
10             bRes = True
11             Exit For
12         End If
13     Next i
14
15     HasTable = bRes
16 End Function
```

In this function, we first call the function ListTables to get the names of
all tables in the database. Then we use a For-Next loop to check whether
the given table name matches one of the names in the database. To test this
function, we use the following sub procedure:

```
 1 Sub TestHasTable()
 2     Dim sFileName As String
 3
 4     sFileName = ThisWorkbook.Path & "\va.accdb"
 5     Connect (sFileName)
```

```
 6
 7      Dim  vRes  As  Variant
 8      vRes  =  ListTables
 9
10      Debug . Print  HasTable ( " Inforce10k " )
11      Debug . Print  HasTable ( " 2008 " )
12
13      Disconnect
14 End  Sub
```

Executing the above test code gives:

```
1 True
2 False
```

The output indicates that table "2008" does not exist in the database.

The following sub procedure is used to delete a table in the database:

```
 1 Sub  DeleteTable ( ByVal  TableName  As  String )
 2      Dim  sSQL  As  String
 3      Dim  dbCmd  As  ADODB . Command
 4
 5      Set  dbCmd  =  New  ADODB . Command
 6      sSQL  =  " DROP  TABLE  "  &  TableName
 7
 8      dbCmd . ActiveConnection  =  dbConnection
 9      dbCmd . CommandText  =  sSQL
10
11      dbCmd . Execute
12
13      Set  dbCmd  =  Nothing
14 End  Sub
```

In this function, we use the ADODB.Command object to execute the SQL query. In Lines 8–9, we specify the properties ActiveConnection and CommandText for the object. In Line 11, we call the method Execute to run the SQL query to delete the table.

———————————————⌒———————————————

Exercise 11.1. Open the database va.accdb and manually create a table named Table1 in the database. Then write a sub procedure named TestDeleteTable to delete the table Table1 from the database.

———————————————⌒———————————————

To create a table in the database, we use the following sub procedure:

```
1  Sub CreateTable(ByRef Data As Range, ByVal TableName
       As String, ByVal PrimaryKey As String)
2      If HasTable(TableName) Then
3          DeleteTable (TableName)
4      End If
5
6      Dim sCmdText As String
7
8      sCmdText = "CREATE TABLE " & TableName & " ("
9
10     Dim i As Integer
11     For i = 1 To Data.Columns.Count
12         Select Case VarType(Data.Cells(2, i).Value)
13             Case vbDouble
14                 sCmdText = sCmdText & "[" & Data.
                       Cells(1, i).Value & "] DOUBLE"
15             Case vbString
16                 sCmdText = sCmdText & "[" & Data.
                       Cells(1, i).Value & "] VARCHAR"
17             Case vbDate
18                 sCmdText = sCmdText & "[" & Data.
                       Cells(1, i).Value & "] DATETIME"
19         End Select
20         If StrComp(Data.Cells(1, i).Value, PrimaryKey
               , vbTextCompare) = 0 Then
21             sCmdText = sCmdText & " PRIMARY KEY"
22         End If
23         If i < Data.Columns.Count Then
24             sCmdText = sCmdText & ", "
25         End If
26     Next i
27     sCmdText = sCmdText & ")"
28
29     Dim dbCmd As ADODB.Command
30     Set dbCmd = New ADODB.Command
31
32     dbCmd.ActiveConnection = dbConnection
33     dbCmd.CommandText = sCmdText
34
35     Debug.Print sCmdText
36     dbCmd.Execute
37     Set dbCmd = Nothing
38 End Sub
```

This sub procedure requires three arguments: a range of data, a table name, and the name of the primary key. We assume that the first row of the range contains the column names of the table. In Lines 8–26, we construct the SQL query for creating the table. The data types are derived from the data in the second row of the range. Here, we only consider three types of data: Double, String, and Date. We also assume that the first column of the range is the primary key, which is a special column designated to uniquely identify all table records. In Lines 28–34, we create an ADODB.Command object to execute query.

Exercise 11.2. Suppose that the range "O2:Q5" contains the following data:

First Name	Last Name	Grade
Courtney	Salazar	A
Carroll	Paul	B
Jane	Taylor	C

Write a sub procedure named TestCreateTable to create a table named Grade in the database based on the range "O2:Q5".

The following sub procedure can be used to save a table in a worksheet to the database:

```
1  Sub ExcelToAccess (ByRef Data As Range , ByVal
       TableName As String)
2      CreateTable Data:=Data , TableName:=TableName
3
4      Dim dbRecordSet As ADODB.Recordset
5      Set dbRecordSet = New ADODB.Recordset
6      dbRecordSet.CursorLocation = adUseServer
7      dbRecordSet.ActiveConnection = dbConnection
8      dbRecordSet.CursorType = adOpenDynamic
9      dbRecordSet.LockType = adLockOptimistic
10     dbRecordSet.Source = TableName
11
12     dbRecordSet.Open
13
14     Dim i As Integer
15     Dim j As Integer
```

```
16    For i = 2 To Data.Rows.Count
17        dbRecordSet.AddNew
18        For j = 1 To Data.Columns.Count
19            dbRecordSet(Data.Cells(1, j).Value).Value
                 = Data.Cells(i, j).Value
20        Next j
21        dbRecordSet.Update
22    Next i
23
24    dbRecordSet.Close
25    Set dbRecordSet = Nothing
26 End Sub
```

In Line 2, we create the table in the database by calling the sub procedure CreateTable. In Lines 4–10, we create and configure an ADODB.Recordset object. In Line 12, we open the record set for appending new records. In Lines 16–22, we use a loop to append the rows in the range to the record set.

———————————◕———————————

Exercise 11.3. Continue from Exercise 11.2 and write a sub procedure named TestExcelToAccess to save the range "O2:Q5" to a table named Grade in the database.

———————————◕———————————

The following function is used to retrieve data from the database:

```
1 Function ExtractRecord(ByVal SQL As String) As
     Variant
2    Dim dbRecordSet As ADODB.Recordset
3    Dim dbCmd As ADODB.Command
4
5    Set dbCmd = New ADODB.Command
6    dbCmd.ActiveConnection = dbConnection
7    dbCmd.CommandText = SQL
8
9    Set dbRecordSet = dbCmd.Execute
10
11   Dim vTmp As Variant
12   vTmp = dbRecordSet.GetRows
13
14   Dim aRes() As String
15   Dim iU1 As Integer
16   Dim iU2 As Integer
```

```
17      Dim i As Integer
18      Dim j As Integer
19      iU1 = UBound(vTmp, 1)
20      iU2 = UBound(vTmp, 2)
21      ReDim aRes(0 To iU1, 0 To iU2 + 1)
22      For j = 0 To iU1 - 1
23          aRes(j, 0) = dbRecordSet.Fields(j).Name
24      Next j
25      For i = 0 To iU2 - 1
26          For j = 0 To iU1 - 1
27              aRes(j, i + 1) = vTmp(j, i)
28          Next j
29      Next i
30
31      dbRecordSet.Close
32      Set dbRecordSet = Nothing
33      Set dbCmd = Nothing
34
35      ExtractRecord = aRes
36  End Function
```

The function requires one argument, which is an SQL query. In Lines 2–12, we create an ADODB.Recordset object and an ADODB.Command object to retrieve data from the database. In Lines 22–29, we convert the record set to a two-dimensional array.

———————————————————◆———————————————————

Exercise 11.4. Suppose that the database va.accdb contains a table named Grade with the data given in Exercise 11.2. Write a sub procedure named TestExtractRecord to retrieve the students whose grade is B.

———————————————————◆———————————————————

11.3.2 MInterface Module

The MInterface module contains VBA macros for the two buttons shown in Figure 11.1. The macro assigned to the first button is as follows:

```
1  Sub Button1_Click()
2      Dim sFileName As String
3      Dim sSQL As String
4      Dim vRes As Variant
5
```

```
6    sFileName = ThisWorkbook.Path & "\" & Range("
        FileName").Value
7    Connect (sFileName)
8
9    sSQL = "SELECT * FROM " & Range("TableName").
        Value & _
10       " WHERE year([issueDate])=" & Range("
            IssueYear").Value & ";"
11   vRes = ExtractRecord(sSQL)
12
13   Disconnect
14
15   Dim ws As Worksheet
16   Set ws = Sheets("Cohort")
17   ws.Cells.ClearContents
18
19   Dim i As Integer
20   Dim j As Integer
21   For i = 0 To UBound(vRes, 2)
22       For j = 0 To UBound(vRes, 1)
23           ws.Cells(i + 1, j + 1).Value = vRes(j, i)
24       Next j
25   Next i
26 End Sub
```

The sub procedure uses three named ranges defined in Table 11.1. In Line 10, the SQL function year is used to get the year part from a date. In Lines 21–25, we write the retrieved contracts to the worksheet "Cohort."

If we click the first button, we see that output as shown in Figure 11.4. From the output, we see that all the contracts have an issue date in 2009.

The macro assigned to the second button is given below:

```
1 Sub Button2_Click()
2     Dim ws As Worksheet
3
4     Set ws = Sheets("Cohort")
5
6     Dim nCol As Integer
7     Dim nRow As Integer
8     nCol = ws.Cells(1, 1).End(xlToRight).Column
9     nRow = ws.Cells(1, 1).End(xlDown).Row
10
11    Dim sTableName As String
12    sTableName = "Cohort" & Range("IssueYear")
```

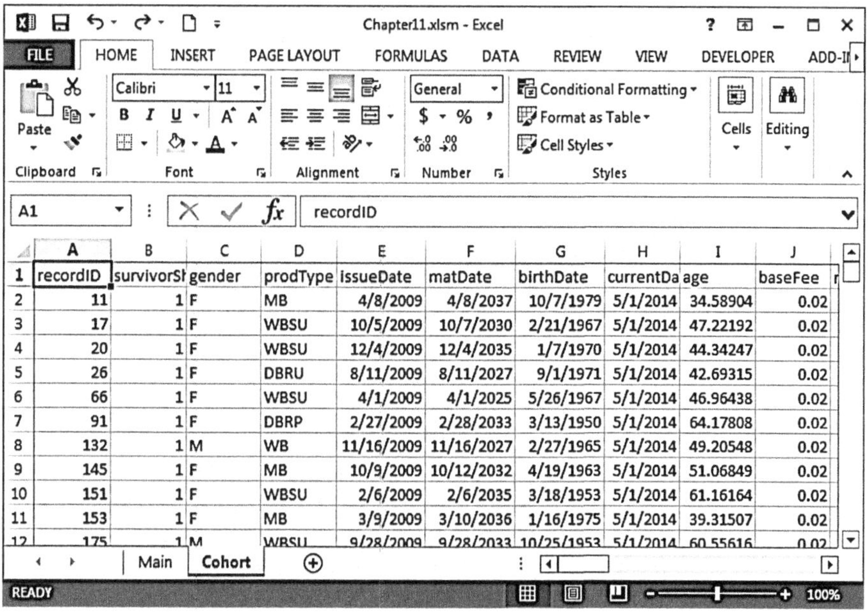

Figure 11.4: Output in the worksheet when the first button is clicked

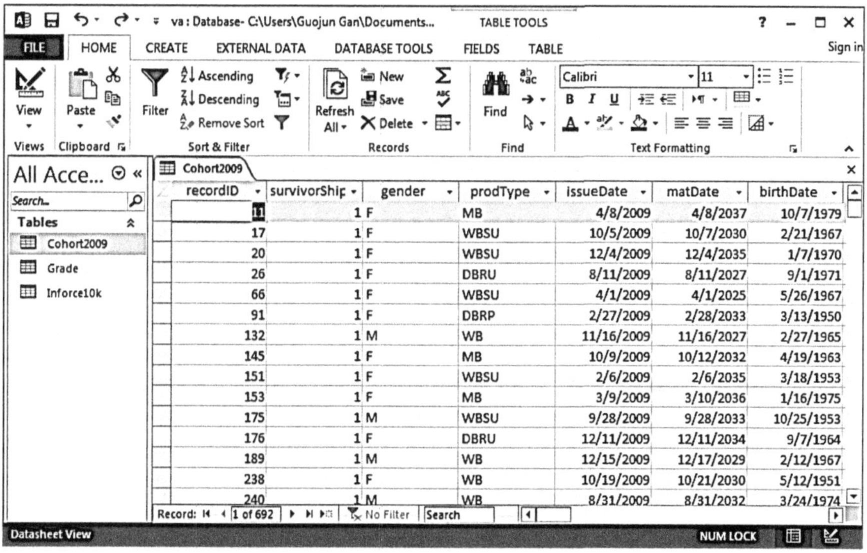

Figure 11.5: New table in the database when the second button is clicked

```
13      Dim rngData As Range
14      Set rngData = ws.Range(ws.Cells(1, 1), ws.Cells(
           nRow, nCol))
15
16      Dim sFileName As String
17      sFileName = ThisWorkbook.Path & "\" & Range("
           FileName").Value
18      Connect (sFileName)
19
20      ExcelToAccess Data:=rngData, TableName:=
           sTableName, PrimaryKey:="recordID"
21
22      Disconnect
23  End Sub
```

In this sub procedure, we just write the table in the "Cohort" worksheet to the database. The name of the table in the database is "Cohort" plus the issue year. If we click the second button, the data in the worksheet "Cohort" will be saved to a new table, as shown in Figure 11.5.

Exercise 11.5. Modify the variable annuity inforce tool as follows:

- Add a new parameter named Product Type to the input area of the interface shown in Figure 11.1.
- Modify the macro assigned to the first button to retrieve the contracts with specified issue year and product type.

How many contracts with "MB" guarantee have an issue date in 2009?

11.4 Summary

In this chapter, we introduced how to connect to a Microsoft Access database in Excel VBA. In particular, we introduced how to retrieve data from an Access database to Excel, how to create tables in Access, and how to transport data in Excel to an Access database. To learn more about the SQL, readers are referred to (Forta, 2012; Tale, 2016).

12

Object-Oriented Programming

In this chapter, we use object-oriented programming techniques to implement the bootstrapping tool introduced in Chapter 8. The interface of the bootstrapping tool to be implemented in this chapter is identical to the one shown in Figure 8.1. However, we implement the bootstrapping tool in an object-oriented way by using classes.

12.1 Introduction

Object-oriented programming (OOP) is a programming paradigm introduced in the 1980s (Urtis, 2015). Since its introduction, OOP has grown in popularity and been supported in many other programming languages, such as C++ (Gan, 2011). It is not surprising that today's Visual Basic for Applications (VBA) programming language also supports OPP.

A computer program developed using the OOP method consists of individual objects, each of which has its own properties and methods. For example, Excel® is an application developed using the OOP method. In Section 2.1, we introduced the Excel® object model. For example, a range is an object in Excel® and contains properties (e.g., value and text) and methods (e.g., select and copy), as described in Section 2.5.

In OOP, the objects of a computer program interact with each other. Each object models an aspect of the problem one tries to solve. In OOP, an object is called an instance of a class, which can be thought as a template of objects. For example, Range is a class in Excel® VBA, and Range("A1") is a Range object. In Section 1.3, we introduced how to define a VBA class.

An object-oriented (OO) VBA program has the following characteristics (Sintes, 2001; Gan, 2011):

(a) Natural. An OO program solves a problem by modeling the problem at a functional level.

(b) Reliable. An OO program consists of well-designed objects. Changing an object will not affect other objects as long as we do not change the interface of the object.

(c) Reusable. The classes created for an OO program can be used in other programs.

(d) Maintainable. Since changing an object does not affect other objects, fixing one part of an OO program will not cause changes in other parts.

Unlike other OOP languages, such as C++ and Java, VBA does not support inheritance. In other words, we cannot define a class by inheriting properties and methods from another class.

12.2   Objects

As we mentioned before, an OO program consists of objects, each of which models an aspect of the problem. To develop an OO bootstrapping tool, we need the following objects: dates, yield curves, holiday calendars, swaps, and a bootstrapping procedure.

Since an object is created from a class, which serves as a template for objects. According to our analysis above, we need classes for dates, yield curves, holiday calendars, swaps, and the bootstrapping procedure. Since the VBA has the Date data type, we do not need to implement our date class. We also do not need to implement a holiday calendar class, because we do not need to keep copies of a calendar object. A holiday calendar is global in the sense that we only need one copy of the holiday calendar. As a result, we can just implement the functions of a holiday calendar in a regular module rather than in a class module. Note that the VBA does not keep separate copies of data in a regular module. However, VBA keeps separate copies of data stored in a class module.

12.3   Implementation

To develop the OO bootstrapping tool, we decide to implement three classes: a yield curve class, a swap class, and a bootstrapper class. These classes allow us to create yield curve objects, swap objects, and bootstrapper objects, respectively.

12.3.1 CZeroCurve Module

We implement a yield curve class in a class module named CZeroCurve. The declarations section of this class module contains the following code:

```
1  Option Explicit
2
3  Dim datCurveDate As Date
4  Dim sYCDCC As String
5  Dim aDate() As Date
6  Dim aDF() As Double
```

In the above code, we declare four module-level variables, which are properties used to store the curve date, the yield curve day count convention, dates, and discount factors at these dates.

The sub procedure used to initialize the module-level variables is implemented as follows:

```
1  Sub Initialize(ByVal CurveDate As Date, ByVal YCDCC
       As String)
2      datCurveDate = CurveDate
3      sYCDCC = YCDCC
4
5      ReDim aDate(0 To 0)
6      ReDim aDF(0 To 0)
7
8      aDate(0) = CurveDate
9      aDF(0) = 1
10 End Sub
```

This sub procedure requires two arguments: the curve date and the yield curve day count convention. The array aDate is initialized to contain one element, which is the curve date. The array aDF is also initialized to contain one element, which is the discount factor at the curve date.

Sometimes we need to make a copy of a CZeroCurve object. To do that, we can use the Clone method implemented as follows:

```
1  Function Clone() As CZeroCurve
2      Dim cZC As CZeroCurve
3      Dim iU As Integer
4      Dim i As Integer
5
6      Set cZC = New CZeroCurve
7      cZC.Initialize CurveDate:=datCurveDate, YCDCC:=
           sYCDCC
```

```
 8      For i = LBound(aDate) + 1 To UBound(aDate)
 9          cZC.PushBack datIn:=aDate(i), dDF:=aDF(i)
10      Next i
11
12      Set Clone = cZC
13 End Function
```

In Line 6 of the above code, we use the New function to create a new object. In Lines 7–10, we copy the data of this object to the new object. This function calls another function called PushBack to copy the dates and the discount factors from this object to a new object called cZC. Note that we use Set to assign the new object to the function name at the end of this function. We have to use Set here because cZC is an object. This method is useful, as it allows us to experiment with a copy of a CZeroCurve object without affecting the object.

The PushBack method is implemented as follows:

```
 1 Sub PushBack(ByVal datIn As Date, ByVal dDF As Double
   )
 2     Dim iU As Integer
 3     iU = UBound(aDate)
 4
 5     If datIn <= aDate(iU) Then
 6         Err.Raise Number:=2002, Source:="CZeroCurve.
               PushBack", Description:="The input date is
               ealier than the last date."
 7     End If
 8
 9     ReDim Preserve aDate(0 To iU + 1)
10     ReDim Preserve aDF(0 To iU + 1)
11     aDate(iU + 1) = datIn
12     aDF(iU + 1) = dDF
13 End Sub
```

The method PushBack is a sub procedure that inserts a date and a discount factor to the end of the arrays aDate and aDF. This method takes two arguments: a date and a real number. The method checks whether the input date is equal to or before the last date in the array aDate. If the input date is equal to or before the last date in the array, the method will raise an error. This method ensures that the dates in the array are sorted in an increasing order.

We implement the raw interpolation method described in Section 8.2 in the following method:

```
1  Function LogLinear(ByVal datD As Date) As Double
2      If datD < datCurveDate Then
3          Err.Raise Number:=2001, Source:="CZeroCurve.
               LogLinear", Description:="Date is ealier
               than the curve date."
4      End If
5      Dim i As Integer
6      Dim iL As Integer
7      Dim iH As Integer
8      iL = LBound(aDate)
9      iH = UBound(aDate)
10     For i = iL To iH
11         If datD = aDate(i) Then
12             LogLinear = aDF(i)
13             Exit Function
14         End If
15     Next i
16
17     Dim dat1 As Date
18     Dim dat2 As Date
19     Dim dDF1 As Double
20     Dim dDF2 As Double
21     Dim bFound As Boolean
22     bFound = False
23     For i = iL To iH - 1
24         If datD > aDate(i) And datD < aDate(i + 1)
               Then
25             dat1 = aDate(i)
26             dat2 = aDate(i + 1)
27             dDF1 = aDF(i)
28             dDF2 = aDF(i + 1)
29             bFound = True
30         End If
31     Next i
32
33     If Not bFound Then
34         dat1 = aDate(iH - 1)
35         dat2 = aDate(iH)
36         dDF1 = aDF(iH - 1)
37         dDF2 = aDF(iH)
38     End If
39
40     Dim dDCF1 As Double
41     Dim dDCF2 As Double
```

```
42    Dim dDCF As Double
43    dDCF1 = TFrac(datCurveDate, dat1, sYCDCC)
44    dDCF2 = TFrac(datCurveDate, dat2, sYCDCC)
45    dDCF = TFrac(datCurveDate, datD, sYCDCC)
46
47    Dim dTemp As Double
48    dTemp = WorksheetFunction.Ln(dDF2) * (dDCF -
          dDCF1) + _
49        WorksheetFunction.Ln(dDF1) * (dDCF2 - dDCF)
50    LogLinear = Exp(dTemp / (dDCF2 - dDCF1))
51 End Function
```

This function takes one argument, which is a date, and returns the discount factor at the input date. If the input date is older than the curve date, the function will raise an error. If the input date matches one of the dates saved in the array aDate, the function just returns the corresponding discount factor saved in the array aDF. Otherwise, the function searches for two dates in aDate that enclose the input date. If the input date is not enclosed in any two dates in aDate, the function will use the last two dates in aDate. Then, the function performs the raw interpolation based on the two dates and the two corresponding discount factors.

The following function allows us to get the dates corresponding to the discount factors:

```
1 Function Dates() As Variant
2     Dates = aDate
3 End Function
```

The method Dates returns the array aDate as a variant. This method allows us to extract the dates saved in the variable aDate, which is private to a CZeroCurve object.

The methods mentioned above allow us to perform necessary actions on a CZeroCurve object. If needed, we can add more methods to the class.

12.3.2 CSwap Module

The module CSwap is a class module implementing an interest rate swap. The declarations section of this class module contains the following code:

```
1 Option Explicit
2
3 Dim sFixFreq As String
4 Dim sFixDCC As String
```

```
 5  Dim sFltFreq As String
 6  Dim sFltDCC As String
 7  Dim sCalendar As String
 8  Dim sBDC As String
 9  Dim iSetDays As Integer
10  Dim datEffective As Date
11  Dim sTenor As String
12  Dim dRate As Double
13
14  Dim datSet As Date
15  Dim vFixPaymentDates As Variant
16  Dim vFltPaymentDates As Variant
```

As we discussed in Section 8.3, an interest rate swap consists of two legs: a fixed leg and a floating leg. The properties defined in Lines 3–12 represent various specifications of an interest rate swap contract. The properties defined in Lines 14–16 hold some convenient values calculated from other properties.

The properties sFixFreq and sFixDCC hold the payment frequency and the day count convention of the fixed leg. Similarly, the properties sFltFreq and sFltDCC hold the payment frequency and the day count convention of the floating leg. In our implementation, we allow the fixed leg and the floating leg of a swap to have different specifications.

The properties sCalendar and sBDC hold the business calendar and the business day count convention of the yield curve. For example, the US yield curve uses the US holiday calendar. The business day count convention tells how to handle the cases when a payment date falls on a holiday or weekend.

The property iSetDays is used to hold the number of settlement days of a swap. For example, the number of settlement days in the US is 2 business days, meaning that a swap transaction is settled 2 business days after the transaction date. The transaction date of a swap is stored in the property datEffective.

The properties sTenor and dRate hold the tenor and the fixed rate of a swap. The value of sTenor can be "1Y," "5Y," etc.

The remaining properties are used to store some values calculated from the specifications of a swap. For example, the property datSet holds the settlement date of the swap. The properties vFixPaymentDates and vFltPaymentDates hold the payment dates of the fixed leg and the floating leg, respectively.

The class CSwap contains only two methods, which are given in the following code:

```
 1 Sub Initialize(ByVal FixFreq As String, ByVal FixDCC
      As String, ByVal FltFreq As String, _
 2      ByVal FltDCC As String, ByVal Calendar As String,
          ByVal BDC As String, _
 3      ByVal SetDays As Integer, ByVal EffectiveDate As
          Date, ByVal Tenor As String, _
 4      ByVal Rate As Double)
 5
 6      sFixFreq = FixFreq
 7      sFixDCC = FixDCC
 8      sFltFreq = FltFreq
 9      sFltDCC = FltDCC
10      sCalendar = Calendar
11      sBDC = BDC
12      iSetDays = SetDays
13      datEffective = EffectiveDate
14      sTenor = Tenor
15      dRate = Rate
16
17      Dim i As Integer
18      datSet = datEffective
19      For i = 1 To iSetDays
20          datSet = RollDate(AddDays(datSet, 1), "Follow
              ", sCalendar)
21      Next i
22
23      vFixPaymentDates = GenerateSchedule(datSet,
          sFixFreq, sTenor, sCalendar, sBDC)
24      vFltPaymentDates = GenerateSchedule(datSet,
          sFltFreq, sTenor, sCalendar, sBDC)
25 End Sub
26
27 Function NPV(ByRef cZC As CZeroCurve) As Double
28      Dim dFixPV As Double
29      Dim dFltPV As Double
30      Dim dDF As Double
31      Dim i As Integer
32      Dim dT As Double
33      Dim dFR As Double
34
35      For i = LBound(vFixPaymentDates) + 1 To UBound(
          vFixPaymentDates)
```

```
36        dDF = cZC.LogLinear(vFixPaymentDates(i))
37        dFixPV = dFixPV + TFrac(vFixPaymontDates(i -
              1), vFixPaymentDates(i), sFixDCC) * dRate
              * dDF
38     Next i
39
40     For i = LBound(vFltPaymentDates) + 1 To UBound(
           vFltPaymentDates)
41        dDF = cZC.LogLinear(vFltPaymentDates(i))
42        dT = TFrac(vFltPaymentDates(i - 1),
              vFltPaymentDates(i), sFltDCC)
43        dFR = (cZC.LogLinear(vFltPaymentDates(i - 1))
              / dDF - 1) / dT
44        dFltPV = dFltPV + dT * dFR * dDF
45     Next i
46
47     NPV = dFixPV - dFltPV
48 End Function
```

The method Initialize is used to initialize an object by setting values of the properties. The payment dates of the fixed leg and the floating leg are created by calling the function GenerateSchedule, which is a function defined in the module MHoliday (see Section 7.6.2).

The method NPV calculates the net present value of a swap. This method takes one argument, which is a reference to a yield curve object. This method first calculates the present value of the fixed cash flows and then calculates the present value of the floating cash flows.

12.3.3 CBootstrapper Module

The bootstrapper class is implemented in the class module CBootstrapper. The declarations section of this class module contains the following code:

```
1  Option Explicit
2
3  ' input
4  Dim aTenor() As String
5  Dim aRate() As Double
6  Dim sFixFreq As String
7  Dim sFixDCC As String
8  Dim sFltFreq As String
9  Dim sFltDCC As String
10 Dim sCalendar As String
11 Dim sBDC As String
```

```
12 Dim datCurveDate As Date
13 Dim iSetDays As Integer
14 Dim sYCDCC As String
15
16 ' settlement date
17 Dim datSet As Date
18
19 ' yield curve
20 Dim cZC As CZeroCurve
```

The properties of the class CBootstrapper are used to store the input and output data. For example, the properties aTenor and aRate stores the tenors and rates of the input swap rates. The properties sFixFreq and sFixDCC represent the payment frequency and the day count convention of the fixed leg. The properties sFltFreq and sFltDCC represent the payment frequency and the day count convention of the floating leg. The properties sCalendar and sBDC hold the business calendar and the business day count convention of the swap. The properties datCurveDate, iSetDays, and sYCDCC store the curve date, the number of settlement days, and the curve's day count convention, respectively.

The remaining two properties are used to store some immediate and output values. For example, the property datSet stores the settlement date of the swaps. For convenience purposes, we save the settlement date into a variable, which can be used later. The property cZC stores the yield curve that is bootstrapped from the input data.

To initialize the properties, we implement the following method:

```
1  Sub Initialize(ByVal FixFreq As String, ByVal FixDCC
       As String, ByVal FltFreq As String, _
2      ByVal FltDCC As String, ByVal Calendar As String,
           ByVal BDC As String, _
3      ByVal SetDays As Integer, ByVal EffectiveDate As
           Date, ByVal YCDCC As String, _
4      ByRef Tenors() As String, ByRef Rates() As Double
           )
5      Dim iL As Integer
6      Dim iU As Integer
7      Dim i As Integer
8
9      iL = LBound(Tenors)
10     iU = UBound(Tenors)
11
12     ReDim aTenor(iL To iU)
```

```
13    ReDim aRate(iL To iU)
14
15    For i = iL To iU
16        aTenor(i) = Tenors(i)
17        aRate(i) = Rates(i)
18    Next i
19
20    sFixFreq = FixFreq
21    sFixDCC = FixDCC
22    sFltFreq = FltFreq
23    sFltDCC = FltDCC
24    sCalendar = Calendar
25    sBDC = BDC
26    iSetDays = SetDays
27    datCurveDate = EffectiveDate
28    sYCDCC = YCDCC
29
30    datSet = EffectiveDate
31    For i = 1 To iSetDays
32        datSet = RollDate(AddDays(datSet, 1), "Follow
              ", sCalendar)
33    Next i
34
35    Set cZC = New CZeroCurve
36    cZC.Initialize CurveDate:=datCurveDate, YCDCC:=
          sYCDCC
37 End Sub
```

The method `Initialize` is used to initialize the properties of an object. The values used to initialize these properties are passed from arguments. Most arguments of this method are ByVal arguments except for the arguments Tenors and Rates, which are ByRef arguments. We have to use the ByRef modifier for Tenors and Rates because it is not possible to pass an array using ByVal.

Newton's method described in Section 8.4 is implemented in the following method:

```
1 Private Sub SolveRate(ByVal iInd As Integer)
2    Dim cIRS As CSwap
3    Dim cZCTmp As CZeroCurve
4    Dim datMaturity As Date
5    Dim dRate As Double
6
7    Dim iNum As Integer
```

```
 8     Dim sPeriod As String
 9
10     ExtractPeriod sIn:=aTenor(iInd), iNum:=iNum,
          sPeriod:=sPeriod
11     datMaturity = OffsetDate(datSet, iNum, sPeriod)
12
13     Set cIRS = New CSwap
14     cIRS.Initialize FixFreq:=sFixFreq, FixDCC:=
          sFixDCC, FltFreq:=sFltFreq, FltDCC:=sFltDCC, _
15         Calendar:=sCalendar, BDC:=sBDC, SetDays:=
              iSetDays, EffectiveDate:=datCurveDate, _
16         Tenor:=aTenor(iInd), Rate:=aRate(iInd)
17
18     Dim dx As Double
19     Dim dx0 As Double
20     Dim dx1 As Double
21     Dim df As Double
22     Dim df0 As Double
23     Dim df1 As Double
24
25     dx0 = Exp(-aRate(iInd) * TFrac(datCurveDate,
          datMaturity, sYCDCC))
26     Set cZCTmp = cZC.Clone
27     cZCTmp.PushBack datIn:=datMaturity, dDF:=dx0
28     df0 = cIRS.NPV(cZCTmp)
29
30     dx1 = dx0 + 0.001
31     Set cZCTmp = cZC.Clone
32     cZCTmp.PushBack datIn:=datMaturity, dDF:=dx1
33     df1 = cIRS.NPV(cZCTmp)
34     Do While Abs(dx1 - dx0) > 0.0000000001
35         dx = dx0 - df0 * (dx1 - dx0) / (df1 - df0)
36         Set cZCTmp = cZC.Clone
37         cZCTmp.PushBack datIn:=datMaturity, dDF:=dx
38         df = cIRS.NPV(cZCTmp)
39
40         dx0 = dx1
41         df0 = df1
42         dx1 = dx
43         df1 = df
44     Loop
45
46     cZC.PushBack datIn:=datMaturity, dDF:=dx
47 End Sub
```

The method `SolveRate` bootstraps a discount factor from a given swap rate. The new discount factor is added to the yield curve.

The following method bootstraps all discount factors:

```vba
1 Sub BuildCurve()
2     Dim iL As Integer
3     Dim iH As Integer
4     Dim i As Integer
5     iL = LBound(aTenor)
6     iH = UBound(aTenor)
7
8     For i = iL To iH
9         SolveRate iInd:=i
10    Next i
11 End Sub
```

The method `BuildCurve` calls the method `SolveRate` repeatedly to bootstrap all the discount factors based on the given swap rates.

To calculate the output specified by users, we implement the following method:

```vba
1 Function CreateOutput(ByVal OutputType As String) As
    Variant
2     Dim aOutDates() As Date
3     Dim i As Integer
4     Dim iL As Integer
5     Dim iH As Integer
6     Dim dT As Double
7     Dim aDate As Variant
8
9     If StrComp(OutputType, "Input", vbTextCompare) =
        0 Then
10        aDate = cZC.Dates
11        iL = LBound(aDate)
12        iH = UBound(aDate)
13        ReDim aOutDates(iL To iH)
14        For i = iL To iH
15            aOutDates(i) = aDate(i)
16        Next i
17    Else
18        aOutDates = GenerateSchedule(datCurveDate,
            OutputType, "30Y", sCalendar, sBDC)
19    End If
20
21    iL = LBound(aOutDates)
```

```
22    iH = UBound(aOutDates)
23
24    ReDim aOut(iL To iH, 0 To 4) As Double
25    For i = iL To iH
26        aOut(i, 0) = CDbl(aOutDates(i))
27        aOut(i, 1) = cZC.LogLinear(aOutDates(i))
28        dT = TFrac(datCurveDate, aOutDates(i), sYCDCC
          )
29        If dT = 0 Then
30            aOut(i, 2) = 0
31        Else
32            aOut(i, 2) = -WorksheetFunction.Ln(aOut(i
              , 1)) / dT
33        End If
34        If i < iH Then
35            aOut(i, 3) = (cZC.LogLinear(aOutDates(i))
                  / cZC.LogLinear(aOutDates(i + 1)) -
                  1) / _
36                TFrac(aOutDates(i), aOutDates(i + 1),
                      sYCDCC)
37        Else
38            aOut(i, 3) = aOut(i - 1, 3)
39        End If
40        aOut(i, 4) = TFrac(datCurveDate, aOutDates(i)
              , sYCDCC)
41    Next i
42    CreateOutput = aOut
43 End Function
```

The method `CreateOutput` calculates the output quantities based on a given output type. If the output type is "Input," then this method calculates the discount factors, forward rates, zero rates, and the year fractions at the maturity dates of the input swap rates. Otherwise, this method calculates these quantities based on a given frequency.

12.3.4   OO Bootstrapping Tool

The OO bootstrapping tool consists of three regular modules and three class modules. The VBA code of the three class modules are shown in Sections 12.3.1, 12.3.2, and 12.3.3, respectively. The three regular modules include `MDate`, `MHoliday`, and `MInterface`. The first two regular modules are implemented in Chapter 7.

The VBA code of the module `MInterface` is given below:

```vba
Sub Button1_Click()
    Dim vRes As Variant
    Dim aTenor() As String
    Dim aRate() As Double
    Dim sFixFreq As String
    Dim sFixDCC As String
    Dim sFltFreq As String
    Dim sFltDCC As String
    Dim sCalendar As String
    Dim sBDC As String
    Dim datCurveDate As Date
    Dim iSetDays As Integer
    Dim sYCDCC As String
    Dim sOutputType As String

    Dim iNumRates As Integer
    Dim i As Integer
    Dim iRows As Integer
    Dim j As Integer

    Dim cBT As CBootstrapper

    iNumRates = Range("Rate").Rows.Count
    ReDim aTenor(1 To iNumRates)
    ReDim aRate(1 To iNumRates)
    With Range("Rate")
        For i = 1 To iNumRates
            aTenor(i) = .Cells(i, 1).Value
            aRate(i) = .Cells(i, 2).Value / 100
        Next i
    End With
    With Range("Param")
        sFixFreq = .Cells(1, 1).Value
        sFixDCC = .Cells(2, 1).Value
        sFltFreq = .Cells(3, 1).Value
        sFltDCC = .Cells(4, 1).Value
        sCalendar = .Cells(5, 1).Value
        sBDC = .Cells(6, 1).Value
        datCurveDate = .Cells(7, 1).Value
        iSetDays = .Cells(8, 1).Value
        sYCDCC = .Cells(9, 1).Value
        sOutputType = .Cells(10, 1).Value
    End With
```

```
45    Set cBT = New CBootstrapper
46    cBT.Initialize FixFreq:=sFixFreq, FixDCC:=sFixDCC
        , FltFreq:=sFltFreq, _
47        FltDCC:=sFltDCC, Calendar:=sCalendar, BDC:=
            sBDC, SetDays:=iSetDays, _
48        EffectiveDate:=datCurveDate, YCDCC:=sYCDCC,
            Tenors:=aTenor, Rates:=aRate
49    cBT.BuildCurve
50    vRes = cBT.CreateOutput(sOutputType)
51
52    iRows = Sheet1.Range("H3").End(xlDown).Row
53    Sheet1.Range(Cells(3, 8), Cells(iRows, 12)).
        ClearContents
54    For i = LBound(vRes, 1) To UBound(vRes, 1)
55        For j = 0 To 4
56            Sheet1.Cells(i + 3, 8 + j).Value = vRes(i
                , j)
57        Next j
58    Next i
59 End Sub
```

The VBA code in the module MInterface is the macro assigned to the button. The sub procedure gets input data from the worksheet, calls other functions to bootstrap the yield curve, and extracts the output data to the worksheet. In Lines 45, a new object of the class CBootstrapper is created. In Line 46, the object is initialized with data from the worksheet. In Lines 49–50, the functions BuildCurve and CreateOutput are called to bootstrap the yield curve and get the output, respectively.

The input and output of the OO bootstrapping tool are identical to those of the bootstrapping tool implemented in Chapter 8. If we click the button of this tool, we see the results shown in Figure 8.1.

Exercise 12.1. In the MInterface module, we see that the three functions Initialize, BuildCurve, and CreateOutput are called in sequence. However, it is possible that some users may not call these functions in sequence. Modify the VBA code of the class CBootstrapper by doing the following:

(a) Add two properties named bInitialized and bCurveBuilt and assign false to them.

(b) Modify the function Initialize by assigning true to the variable bInitalized at the end of this function

(c) Modify the function BuildCurve by checking whether bInitialized is true. If bInitialized is false, raise an error. Also assign true to the variable bCurveBuilt at the end of this function.

(d) Modify the function CreateOutput by checking whether bCurveBuilt is true. If bCurveBuilt is false, raise an error.

Exercise 12.2. In Exercise 12.1, you are asked to modify the code of the class CBootstrapper to address the issue that the three functions have to be called in sequence. However, users still need to call the three functions explicitly. We can modify the code such that users only need to call the functions Initialize and CreateOutput, which calls the other function automatically, if necessary. Modify the VBA code of the class CBootstrapper and the module MInterface as follows:

(a) Add two properties named bInitialized and bCurveBuilt and assign false to them.

(b) Modify the function Initialize by assigning true to the variable bInitalized at the end of this function.

(c) Modify the function BuildCurve by checking whether bInitialized is true. If bInitialized is false, raise an error. Also assign true to the variable bCurveBuilt at the end of this function.

(d) Modify the function CreateOutput by checking whether bCurveBuilt is true. If bCurveBuilt is false, call the function BuildCurve.

(e) Modify the function Button1_Click in the module MInterface by removing the function call cBT.BuildCurve.

12.4 Summary

In this chapter, we implemented a bootstrapping tool using OO techniques. We implemented three classes, each of which models an aspect of the bootstrapping algorithm. In particular, we implemented a yield curve class, a swap class, and a bootstrapper class. The advantage of using classes is that VBA keeps a copy of the data contained in an object. By using classes, we can create multiple objects of a class to contain different copies of data. For more information about class modules, readers are referred to Walkenbach (2013a).

A

Solutions to Selected Exercises

A.1 Introduction to VBA

Solution 1.2. We can create the module by right clicking the Modules folder in the Project window. We can change the module name through the Properties window.

Solution 1.4. The recorded macro looks like:

```
 1  Sub Macro2()
 2  '
 3  ' Macro2 Macro
 4  '
 5
 6  '
 7      Range("A1:B2").Select
 8      Selection.Copy
 9      Range("F1").Select
10      Selection.PasteSpecial Paste:=xlPasteValues,
            Operation:=xlNone, SkipBlanks _
11          :=False, Transpose:=False
12      Application.CutCopyMode = False
13      Selection.Style = "Percent"
14      Selection.NumberFormat = "0.0%"
15      Selection.NumberFormat = "0.00%"
16  End Sub
```

A.2 Excel® Objects

Solution 2.2. We can implement the sub procedure as follows:

```
 1  Sub CountRowColumn()
 2      Debug.Print Rows.Count
```

```
3|     Debug.Print Columns.Count
4| End Sub
```

Executing the above sub procedure gives the following output:

```
1|   1048576
2|   16384
```

Solution 2.4. We can write the sub procedure as follows:

```
1| Sub InputBoxDemo()
2|     Debug.Print Application.InputBox(prompt:="Input a
         number", Type:=1) + 5
3| End Sub
```

Solution 2.6. The first and second statements are true. The third statement is false. If the Visual Basic code is part of an add-in, ThisWorkbook won't return the active workbook. In this case, ActiveWorkbook is the workbook calling the add-in, whereas ThisWorkbook returns the add-in workbook.

Solution 2.8. The Index property is read-only. We cannot assign a value to it.

Solution 2.10. We can find out the address using the following code:

```
1| Sub FindAddress2()
2|     Debug.Print Cells(20, 51).Address
3| End Sub
```

Executing the sub procedure gives "*AY*20."

Solution 2.12. The output is:

```
1| $D$4:$F$5
2| $D$4:$F$5
```

Solution 2.14. We can write the sub procedure as follows:

```
1| Sub CopyTable()
2|     Workbooks("USLifeTable1999-2001Male.xls").Sheets
         (1).Range("A25:B134").Copy
3|     ThisWorkbook.Sheets(1).Range("F1").PasteSpecial
         xlPasteValues
4| End Sub
```

Solution 2.16. We can calculate these quantities as follows:

```
1  Sub WorksheetFunctionDemo3()
2      Debug.Print WorksheetFunction.Pi * 2
3      Debug.Print WorksheetFunction.NormDist(1.5, 0, 1,
           True)
4      Debug.Print WorksheetFunction.NormInv(0.025, 0,
           1)
5  End Sub
```

Executing the code produces the following output:

```
1  6.28318530717959
2   0.933192798731142
3  -1.95996398454005
```

A.3 Variables, Data Types, and Scopes

Solution 3.2. We need to use the Set statement to assign an object to an Object variable.

Solution 3.4. The code does not contain any errors.

Solution 3.6. We can write the sub procedure as follows:

```
1  Sub WorksheetDemo6()
2     Dim ws As Worksheet
3     Set ws = Worksheets.Add
4     ws.Name = "SheetB"
5     Debug.Print ActiveSheet.Name
6  End Sub
```

Solution 3.8. The base can only be 0 or 1.

Solution 3.10. We can write the sub procedure as follows:

```
1  Sub ArrayDemo7()
2      Dim arM(1 To 2, 1 To 2, 1 To 2) As Integer
3
4      arM(1, 1, 1) = 3
5      arM(1, 1, 2) = 4
6      arM(1, 2, 1) = 4
7      arM(1, 2, 2) = 5
8      arM(2, 1, 1) = 4
9      arM(2, 1, 2) = 5
```

```
10      arM(2, 2, 1) = 5
11      arM(2, 2, 2) = 6
12
13      Debug.Print arM(1, 1, 2), arM(1, 2, 1)
14      Debug.Print arM(1, 2, 1), arM(1, 2, 2)
15      Debug.Print arM(2, 2, 1), arM(2, 2, 2)
16      Debug.Print arM(2, 2, 1), arM(2, 2, 2)
17 End Sub
```

Executing the code gives the following output:

```
1   4                 4
2   4                 5
3   5                 6
4   5                 6
```

Solution 3.12. Objects cannot be defined as constants. Defining constants also requires constant expression. In this procedure, conBook, conCell, and conCellsInRow are not valid.

Solution 3.14. We cannot assign new values to constants.

Solution 3.16. We can write the sub procedure as follows:

```
1 Sub StringDemo6()
2     Dim arV As Variant
3     Dim arWord() As String
4     Dim arCount() As Integer
5     Dim i As Integer
6     Dim j As Integer
7     Dim intMax As Integer
8     Dim iL As Integer
9     Dim iU As Integer
10     Dim bFound As Boolean
11     Dim iInd As Integer
12
13     arV = RandomWords
14     iL = LBound(arV)
15     iU = UBound(arV)
16
17     ReDim arWord(1 To 1)
18     ReDim arCount(1 To 1)
19     arWord(1) = arV(iL)
20     arCount(1) = 1
21
22     For i = iL + 1 To iU
23         bFound = False
```

```
24        For j = 1 To UBound(arWord)
25            If StrComp(arV(i), arWord(j),
                 vbTextCompare) = 0 Then
26                bFound = True
27                iInd = j
28            End If
29        Next j
30
31        If bFound Then
32            arCount(iInd) = arCount(iInd) + 1
33        Else
34            ReDim Preserve arWord(1 To UBound(arWord)
                 + 1)
35            ReDim Preserve arCount(1 To UBound(
                 arCount) + 1)
36            arWord(UBound(arWord)) = arV(i)
37            arCount(UBound(arCount)) = 1
38        End If
39    Next i
40
41    intMax = 0
42    For i = 1 To UBound(arWord)
43        If arCount(i) > intMax Then
44            intMax = arCount(i)
45        End If
46    Next i
47
48    Dim dSum As Double
49    For i = 1 To UBound(arWord)
50        If arCount(i) = intMax Then
51            Debug.Print arWord(i); arCount(i)
52        End If
53        dSum = dSum + arCount(i)
54    Next i
55
56    Debug.Print dSum
57 End Sub
```

Executing the above code produces the following output:

```
1 bcbd 6
2  1000
```

Solution 3.18. We can write the sub procedure as follows:

```
1  Sub StringDemo11()
2      Dim strA As String
3      Dim strB As String
4      Dim intA As Integer
5      strA = "This car is beautiful. That car is small.
          Your car is big."
6
7      intA = InStr(strA, "car")
8      strB = Mid(strA, 1, intA) & Replace(strA, "car",
          "house", intA + 1, 1)
9      Debug.Print strB
10     Debug.Print Len(strA)
11     Debug.Print Len(strB)
12 End Sub
```

Solution 3.20. We can write the sub procedure as follows:

```
1  Sub StringDemo14()
2      Dim strA As String
3      strA = "1.1,0.2,3.14,4.2,0.5,4.8,1.3,6.2,1"
4
5      Dim arB() As String
6      Dim arC() As Double
7
8      arB = Split(strA, ",")
9      Dim i As Integer
10
11     ReDim arC(LBound(arB) To UBound(arB))
12     For i = LBound(arB) To UBound(arB)
13         arC(i) = CDbl(arB(i))
14         Debug.Print arC(i),
15     Next i
16 End Sub
```

Solution 3.22. We can write the sub procedure as follows:

```
1  Sub DateDemo6()
2      Dim datBeg As Date
3      Dim datEnd As Date
4      Dim arA() As Date
5      Dim intW As Integer
6      Dim i As Integer
7      datBeg = #1/1/2016#
8      datEnd = #5/31/2016#
9      intW = DateDiff("w", datBeg, datEnd)
```

```
10    ReDim arA(1 To intW + 1)
11    For i = 1 To intW + 1
12        arA(i) = DateAdd("ww", i - 1, datBeg)
13        Debug.Print arA(i)
14    Next i
15 End Sub
```

The output is:

```
1  1/1/2016
2  1/8/2016
3  1/15/2016
4  1/22/2016
5  1/29/2016
6  2/5/2016
7  2/12/2016
8  2/19/2016
9  2/26/2016
10 3/4/2016
11 3/11/2016
12 3/18/2016
13 3/25/2016
14 4/1/2016
15 4/8/2016
16 4/15/2016
17 4/22/2016
18 4/29/2016
19 5/6/2016
20 5/13/2016
21 5/20/2016
22 5/27/2016
```

Solution 3.24. We can write the function as follows:

```
1  Function IsMLK(dat As Date)
2      Dim m As Integer
3      Dim d As Integer
4      Dim w As Integer
5      m = Month(dat)
6      d = Day(dat)
7      w = Weekday(dat)
8
9      If m = 1 And w = 2 And (d >= 15 And d <= 21) Then
10         IsMLK = True
11     Else
```

```
12        IsMLK = False
13     End If
14 End Function
```

The output is

```
1 True
2 True
3 False
```

A.4   Operators and Control Structures

Solution 4.2. The output is 2016.

Solution 4.4. We will get a type mismatch error. The string "abc" cannot be converted to a number.

Solution 4.6. The output is

```
1 True
2 False
```

Solution 4.8. It is an infinite loop. Executing the code may freeze your computer.

Solution 4.10. We can write the sub procedure as follows:

```
1 Sub IfThenDemo4()
2     Dim intDay As Integer
3     Dim intM As Integer
4     Dim intWeekDay As Integer
5
6     intM = Month(Date)
7     intDay = Day(Date)
8     intWeekDay = Weekday(Date)
9
10    ' Labor Day is the first Monday in September
11    If intM = 9 And intDay < 8 And intWeekDay = 2
          Then
12        Debug.Print "Today is Labor day"
13    Else
14        Debug.Print "Today is not Labor day"
15    End If
16 End Sub
```

Solution 4.12. We can write the sub procedure as follows:

```
1  Sub SelectDemo4()
2      Select Case VarType(Range("A1").Value)
3          Case vbDouble
4              Debug.Print "Double"
5          Case vbString
6              Debug.Print "String"
7          Case vbCurrency
8              Debug.Print "Currency"
9          Case vbBoolean
10             Debug.Print "Boolean"
11         Case Else
12             Debug.Print "Other"
13     End Select
14 End Sub
```

Solution 4.14. We can calculate the sum as follows:

```
1  Sub ForNextDemo3()
2      Dim dSum As Double
3      Dim n As Integer
4
5      For n = 10 To 100
6          dSum = dSum + n ^ 2 + 3 * n ^ 3
7      Next n
8
9      Debug.Print dSum
10 End Sub
```

Executing the code gives 76839490.

Solution 4.16. We can create the array as follows:

```
1  Sub ForNextDemo6()
2      Dim vA(1 To 30) As Double
3      Dim n As Integer
4
5      For n = LBound(vA) To UBound(vA) Step 3
6          vA(n) = 2
7          vA(n + 1) = 1
8          vA(n + 2) = 3
9      Next n
10
11     Debug.Print vA(1); vA(2); vA(3); vA(4); vA(5); vA
           (6)
12 End Sub
```

Solution 4.18. We can write the sub procedure as follows:

```
Sub ForNextDemo9()
    Dim mA(1 To 100, 1 To 100) As Integer
    Dim i As Integer
    Dim j As Integer

    For i = 1 To 100
        For j = 1 To 100
            mA(i, j) = i + j - 2
            ActiveSheet.Cells(i, j) = mA(i, j)
        Next j
    Next i
End Sub
```

Solution 4.20. The For-Next statements are not nested properly.

Solution 4.22. The loop is infinite and will not stop.

A.5 Functions, Events, and File IO

Solution 5.2. We can write the VBA function as follows:

```
Function DNorm(x As Double, mu As Double, sigma As
    Double)
    Dim dPi As Double
    dPi = WorksheetFunction.Pi
    DNorm = Exp(-(x - mu) ^ 2 / (2 * sigma ^ 2)) / (
        Sqr(2 * dPi) * sigma)
End Function
```

The output of the sub procedure is

```
0.398942280401433
0.106482668507451
0.398942280401433
0.106482668507451
```

Solution 5.4. We can write the function procedure as follows:

```
Function Factorial(n As Long)
    If n = 0 Then
```

```
3        Factorial = 1
4    Else
5        Factorial = n * Factorial(n - 1)
6    End If
7 End Function
```

The output of the sub procedure is

```
1  1
2  3628800
3  2.43290200817664E+18
```

Solution 5.6. The first function always returns the same random numbers. However, the second function returns different random numbers for different calls.

Solution 5.8. We can add event-handlers for the NewChart and New-Sheet events as follows:

```
1 Private Sub Workbook_NewChart(ByVal Ch As Chart)
2     If ThisWorkbook.Sheets.Count >= 5 Then
3         Application.DisplayAlerts = False
4         Ch.Delete
5         Application.DisplayAlerts = True
6     End If
7 End Sub
8
9 Private Sub Workbook_NewSheet(ByVal Sh As Object)
10    If Sheets.Count > 5 Then
11        Application.DisplayAlerts = False
12        Sh.Delete
13        Application.DisplayAlerts = True
14    End If
15 End Sub
```

Solution 5.10. We can add an event-handler for the Active event of the worksheet as follows:

```
1 Private Sub Worksheet_Activate()
2     MsgBox ActiveSheet.Name
3 End Sub
```

Solution 5.12. We can write the sub procedure as follows:

```
1 Sub FileDemo6()
2     Dim strFile As String
```

```
3  |    Dim strV() As String
4  |    Dim j As Integer
5  |    Dim n As Integer
6  |    strFile = ThisWorkbook.Path & ":words.txt"
7  |
8  |    n = 1
9  |    Open strFile For Input As #1
10 |    Do Until EOF(1)
11 |        Line Input #1, Line
12 |        strV = Split(Line, " ")
13 |        For j = 0 To UBound(strV)
14 |            Cells(n, j + 1).Value = strV(j)
15 |        Next j
16 |        n = n + 1
17 |    Loop
18 |    Close #1
19 | End Sub
```

A.6   Error Handling and Debugging

Solution 6.2. The code contains runtime errors. For example, when the active sheet is a chart sheet or the active sheet is protected, the code will raise runtime errors.

Solution 6.4. We use the watch window to find out the value at the 900th iteration. The value is 0.68491667509079.

A.7   Generating Payment Schedules

Solution 7.2. The adjusted dates are given by the following:

Payment Date	Following	Modified Following	Proceeding
January 16, 2016	January 19	January 19	January 15
May 30, 2016	May 31	May 31	May 27
November 25, 2016	November 28	November 28	November 24

Solution 7.4. We can implement the function as follows:

```
 1 Function GetHolidayNY(ByVal lY As Long, ByVal
      sHoliday As String) As Variant
 2     Dim adHoliday(1 To 2) As Date
 3     Dim datHoliday As Date
 4     Dim datObserved As Date
 5     Dim iWD As Integer
 6
 7     Select Case sHoliday
 8         Case "NewYear"
 9             datHoliday = DateSerial(lY, 1, 1)
10             If Weekday(datHoliday) = vbSaturday Then
11                 datObserved = DateSerial(lY - 1, 12,
                      31)
12             ElseIf Weekday(datHoliday) = vbSunday
                  Then
13                 datObserved = DateSerial(lY, 1, 2)
14             Else
15                 datObserved = datHoliday
16             End If
17         Case "MLK"
18             datHoliday = DateSerial(lY, 1, 1)
19             iWD = Weekday(datHoliday)
20             If iWD <= vbMonday Then
21                 datHoliday = AddDays(datHoliday,
                      vbMonday - iWD + 14)
22             Else
23                 datHoliday = AddDays(datHoliday,
                      vbMonday - iWD + 21)
24             End If
25             datObserved = datHoliday
26         Case "President"
27             datHoliday = DateSerial(lY, 2, 1)
28             iWD = Weekday(datHoliday)
29             If iWD <= vbMonday Then
30                 datHoliday = AddDays(datHoliday,
                      vbMonday - iWD + 14)
31             Else
32                 datHoliday = AddDays(datHoliday,
                      vbMonday - iWD + 21)
33             End If
34             datObserved = datHoliday
35         Case "Memorial"
36             datHoliday = DateSerial(lY, 5, 31)
37             iWD = Weekday(datHoliday)
```

```
38          If iWD = vbSunday Then
39              datHoliday = AddDays(datHoliday, -6)
40          Else
41              datHoliday = AddDays(datHoliday,
                    vbMonday - iWD)
42          End If
43          datObserved = datHoliday
44      Case "Independence"
45          datHoliday = DateSerial(1Y, 7, 4)
46          If Weekday(datHoliday) = vbSaturday Then
47              datObserved = DateSerial(1Y, 7, 3)
48          ElseIf Weekday(datHoliday) = vbSunday
                Then
49              datObserved = DateSerial(1Y, 7, 5)
50          Else
51              datObserved = datHoliday
52          End If
53      Case "Labor"
54          datHoliday = DateSerial(1Y, 9, 1)
55          iWD = Weekday(datHoliday)
56          If iWD <= vbMonday Then
57              datHoliday = AddDays(datHoliday,
                    vbMonday - iWD)
58          Else
59              datHoliday = AddDays(datHoliday,
                    vbMonday - iWD + 7)
60          End If
61          datObserved = datHoliday
62      Case "Columbus"
63          datHoliday = DateSerial(1Y, 10, 1)
64          iWD = Weekday(datHoliday)
65          If iWD <= vbMonday Then
66              datHoliday = AddDays(datHoliday,
                    vbMonday - iWD + 7)
67          Else
68              datHoliday = AddDays(datHoliday,
                    vbMonday - iWD + 14)
69          End If
70          datObserved = datHoliday
71      Case "Veterans"
72          datHoliday = DateSerial(1Y, 11, 11)
73          If Weekday(datHoliday) = vbSaturday Then
74              datObserved = DateSerial(1Y, 11, 10)
75          ElseIf Weekday(datHoliday) = vbSunday
                Then
```

```
 76                      datObserved = DateSerial(1Y, 11, 12)
 77                  Else
 78                      datObserved = datHoliday
 79                  End If
 80              Case "Thanksgiving"
 81                  datHoliday = DateSerial(1Y, 11, 1)
 82                  iWD = Weekday(datHoliday)
 83                  If iWD <= vbThursday Then
 84                      datHoliday = AddDays(datHoliday,
                            vbThursday - iWD + 21)
 85                  Else
 86                      datHoliday = AddDays(datHoliday,
                            vbThursday - iWD + 28)
 87                  End If
 88                  datObserved = datHoliday
 89              Case "Christmas"
 90                  datHoliday = DateSerial(1Y, 12, 25)
 91                  If Weekday(datHoliday) = vbSaturday Then
 92                      datObserved = DateSerial(1Y, 12, 24)
 93                  ElseIf Weekday(datHoliday) = vbSunday
                        Then
 94                      datObserved = DateSerial(1Y, 12, 26)
 95                  Else
 96                      datObserved = datHoliday
 97                  End If
 98              Case Else
 99                  Err.Raise Number:=1008, Source:="MHoliday
                        .GetHoliday", Description:="Unknown
                        holiday."
100          End Select
101
102          adHoliday(1) = datHoliday
103          adHoliday(2) = datObserved
104          GetHolidayNY = adHoliday
105      End Function
```

Solution 7.6. We can implement the function as follows:

```
1   Function EasterSunday(ByVal 1Y As Long) As Date
2       Dim 1N As Long
3       Dim 1G As Long
4       Dim 1E As Long
5       Dim 1H As Long
6       Dim 1SOL As Long
7       Dim 1LUN As Long
```

```
 8     Dim lV As Long
 9     Dim lC As Long
10     Dim lR As Long
11     Dim lS As Long
12
13     lN = 7 - (lY + lY \ 4 - lY \ 100 + lY \ 400 - 1)
           Mod 7
14     lG = 1 + lY Mod 19
15     lE = (11 * lG - 10) Mod 30
16     lH = lY \ 100
17     lSOL = lH - lH \ 4 - 12
18     lLUN = (lH - 15 - (lH - 17) \ 25) \ 3
19     lV = lE \ 24 - lE \ 25 + (lG \ 12) * (lE \ 25 -
           lE \ 26)
20     lE = (11 * lG - 10) Mod 30 - (lSOL - lLUN) Mod 30
           + lV
21     If lE < 24 Then
22         lR = 45 - lE
23     Else
24         lR = 75 - lE
25     End If
26     lC = 1 + (lR + 2) Mod 7
27     lS = lR + (7 + lN - lC) Mod 7
28
29     EasterSunday = DateSerial(lY, 3, lS)
30 End Function
```

A.8   Bootstrapping Yield Curves

Solution 8.2. We can write the function PvSwap as follows:

```
1 Function PvSwap(ByVal sTenor As String, ByVal dRate
      As Double) As Double
2     Dim iNum As Integer
3     Dim sPeriod As String
4     Dim i As Integer
5     Dim dDF As Double
6
7     Dim vFixDates As Variant
8     Dim dFixPV As Double
```

```
 9   ExtractPeriod sIn:=sTenor, iNum:=iNum, sPeriod:=
         sPeriod
10   ' calculate fixed leg cash flows and present
         value
11   vFixDates = GenerateSchedule(datSet, sFixFreq,
         sTenor, sCalendar, sBDC)
12   For i = LBound(vFixDates) + 1 To UBound(vFixDates
         )
13       dDF = LogLinear(vFixDates(i))
14       dFixPV = dFixPV + TFrac(vFixDates(i - 1),
             vFixDates(i), sFixDCC) * dRate * dDF
15   Next i
16
17   PvSwap = dFixPV - LogLinear(datSet) + LogLinear(
         vFixDates(UBound(vFixDates)))
18 End Function
```

Solution 8.4. We can modify the sub procedure as follows:

```
 1 Sub Button1_Click()
 2   Dim vRes As Variant
 3   Initialize rngRate:=Range("Rate"), rngParam:=
         Range("Param")
 4   CheckInput
 5   BuildCurve
 6   vRes = CreateOutput
 7
 8   Dim iRows As Integer
 9   Dim i As Integer
10   Dim j As Integer
11   iRows = Sheet1.Range("H3").End(xlDown).Row
12   Sheet1.Range(Cells(3, 8), Cells(iRows, 12)).
         ClearContents
13   For i = LBound(vRes, 1) To UBound(vRes, 1)
14       For j = 0 To 4
15           Sheet1.Cells(i + 3, 8 + j).Value = vRes(i
                 , j)
16       Next j
17   Next i
18
19   ' plot the zero rates and forward rates
20   Dim bTag As Boolean
21   Dim csPlot As Chart
22   bTag = False
23   For i = 1 To Charts.Count
```

```
24        If StrComp(Charts(i).Name, "Plot") = 0 Then
25            Set csPlot = Charts(i)
26            bTag = True
27        End If
28    Next i
29    If Not bTag Then
30        Set csPlot = Charts.Add(after:=ActiveSheet)
31        csPlot.Name = "Plot"
32    End If
33
34    Dim seriesA As Series
35    For Each seriesA In csPlot.SeriesCollection
36        seriesA.Delete
37    Next seriesA
38
39    Set seriesA = csPlot.SeriesCollection.NewSeries
40    seriesA.XValues = "=Curve!$L$3:$L$" & CStr(UBound
          (vRes, 1) + 3)
41    seriesA.Values = "=Curve!$J$3:$J$" & CStr(UBound(
          vRes, 1) + 3)
42    seriesA.ChartType = xlXYScatterLines
43    seriesA.Name = "Zero Rate"
44    Set seriesA = csPlot.SeriesCollection.NewSeries
45    seriesA.XValues = "=Curve!$L$3:$L$" & CStr(UBound
          (vRes, 1) + 3)
46    seriesA.Values = "=Curve!$K$3:$K$" & CStr(UBound(
          vRes, 1) + 3)
47    seriesA.ChartType = xlXYScatterLines
48    seriesA.Name = "Forward Rate"
49
50 End Sub
```

A.9   Generating Risk-Neutral Scenarios

Solution 9.2. We can revise the sub procedure as follows:

```
1 Sub Button1_Click()
2    Dim rngParam As Range
3    Dim rngFirstFR As Range
4
5    Set rngParam = Sheet1.Range("Param")
```

```
6     Set rngFirstFR = Sheet1.Range("FirstFR")
7
8     Dim aFR() As Double
9     Dim lFirstRow As Long
10    Dim lLastRow As Long
11
12    lFirstRow = rngFirstFR.Cells(1, 1).Row
13    lLastRow = rngFirstFR.Cells(1, 1).End(xlDown).Row
14
15    ReDim aFR(1 To lLastRow - lFirstRow + 1)
16    Dim i As Long
17    For i = 1 To lLastRow - lFirstRow + 1
18        aFR(i) = rngFirstFR.Cells(1, 1).Offset(i - 1,
              0).Value
19    Next i
20
21    MGenerator.Initialize Sigma:=rngParam.Cells(1, 1)
         .Value, _
22        Horizon:=rngParam.Cells(2, 1).Value, _
23        TimeStep:=rngParam.Cells(3, 1).Value, _
24        NumPath:=rngParam.Cells(4, 1).Value, _
25        Seed:=rngParam.Cells(5, 1).Value, _
26        Precision:=rngParam.Cells(6, 1).Value, _
27        FR:=aFR
28
29    Dim vSce As Variant
30    vSce = MGenerator.Generate
31
32    ' write scenarios to Sheet2
33    Sheet2.Cells.ClearContents
34    Dim j As Long
35    For i = 1 To UBound(vSce, 1)
36        For j = 1 To UBound(vSce, 2)
37            Sheet2.Cells(i, j).Value = vSce(i, j)
38        Next j
39    Next i
40
41    ' calculate summary statistics
42    Dim lRow As Long
43    Dim lCol As Long
44    lRow = UBound(vSce, 1)
45    lCol = UBound(vSce, 2)
46    Dim aTmp() As Double
47    ReDim aTmp(1 To lRow)
48
```

```
49    Dim dDelta As Double
50    dDelta = MGenerator.GetDelta
51    For j = 1 To lCol
52        For i = 1 To lRow
53            aTmp(i) = WorksheetFunction.lN(vSce(i, j)
                  )
54        Next i
55
56        Sheet2.Cells(lRow + 2, j).Value =
              WorksheetFunction.Average(aTmp) / dDelta
57        Sheet2.Cells(lRow + 3, j).Value =
              WorksheetFunction.StDev(aTmp) / Sqr(dDelta
              )
58
59    Next j
60 End Sub
```

The function GetDelta is a function of the module MGenerator and is implemented as

```
1 Function GetDelta() As Double
2     GetDelta = CalDelta(sTimeStep)
3 End Function
```

A.10   Valuing a GMDB

Solution 10.2. We can write the test procedure as follows:

```
1 Sub TestInitialize()
2     Dim rngMTMale As Range
3     Dim rngMTFemale As Range
4
5     Set rngMTMale = Range("MTMale")
6     Set rngMTFemale = Range("MTFemale")
7
8     Initialize rngMTMale:=rngMTMale, rngMTFemale:=
          rngMTFemale
9
10        PrintInfo
11 End Sub
```

Solution 10.4. We can implement the test procedure as follows:

```
1  Sub TestProject()
2      Dim bFound As Boolean
3      bFound = False
4      Dim i As Integer
5      For i = 1 To Sheets.Count
6          If StrComp(Sheets(i).Name, "Temp") = 0 Then
7              bFound = True
8              Exit For
9          End If
10     Next i
11     Dim ws As Worksheet
12     If Not bFound Then
13         Set ws = Worksheets.Add
14         ws.Name = "Temp"
15     End If
16
17     MLifeTable.Initialize rngMTMale:=Range("MTMale"), _
           rngMTFemale:=Range("MTFemale")
18     With Range("Param")
19         MGMDB.InitializeContract _
20             Gender:=.Cells(1, 1).Value, _
21             DOB:=.Cells(2, 1).Value, _
22             Premium:=.Cells(3, 1).Value, _
23             MEFee:=.Cells(4, 1).Value, _
24             FundFee:=.Cells(5, 1).Value, _
25             IE:=.Cells(6, 1).Value, _
26             OE:=.Cells(7, 1).Value, _
27             Term:=.Cells(8, 1).Value, _
28             Feature:=.Cells(9, 1).Value, _
29             LapseRate:=.Cells(10, 1).Value, _
30             ValDate:=.Cells(11, 1).Value, _
31             NumSce:=.Cells(12, 1).Value
32     End With
33
34     MGMDB.InitializeScenario Sce:=Sheets("Base"), FR _
           :=Range("FRBase")
35
36     Project dShock:=0, ind:=1
37     Set ws = Sheets("Temp")
38     ws.Cells.ClearContents
39     ws.Cells(1, 1).Value = "ap"
40     ws.Cells(2, 1).Value = "aq"
41     ws.Cells(3, 1).Value = "ad"
```

```
42    ws.Cells(4, 1).Value = "aL"
43    ws.Cells(5, 1).Value = "aA"
44    ws.Cells(6, 1).Value = "aRC"
45    ws.Cells(7, 1).Value = "aGD"
46    ws.Cells(8, 1).Value = "aDB"
47    For i = 0 To 360
48        ws.Cells(1, i + 2).Value = ap(i)
49        If i > 0 Then
50        ws.Cells(2, i + 2).Value = aq(i)
51        End If
52        ws.Cells(3, i + 2).Value = ad(i)
53        ws.Cells(4, i + 2).Value = aL(i)
54        ws.Cells(5, i + 2).Value = aA(i)
55        ws.Cells(6, i + 2).Value = aRC(i)
56        ws.Cells(7, i + 2).Value = aGD(i)
57        ws.Cells(8, i + 2).Value = aDB(i)
58    Next i
59 End Sub
```

Solution 10.6. We can modify the sub procedure as follows:

```
1  Sub Button1_Click()
2      Dim dFMVBase As Double
3      Dim dFMVEU As Double
4      Dim dFMVED As Double
5      Dim dFMVIU As Double
6      Dim DFMVID As Double
7
8      Dim dStartTime As Double
9      dStartTime = Timer
10
11     MLifeTable.Initialize rngMTMale:=Range("MTMale"),
           rngMTFemale:=Range("MTFemale")
12     With Range("Param")
13        MGMDB.InitializeContract _
14            Gender:=.Cells(1, 1).Value, _
15            DOB:=.Cells(2, 1).Value, _
16            Premium:=.Cells(3, 1).Value, _
17            MEFee:=.Cells(4, 1).Value, _
18            FundFee:=.Cells(5, 1).Value, _
19            IE:=.Cells(6, 1).Value, _
20            OE:=.Cells(7, 1).Value, _
21            Term:=.Cells(8, 1).Value, _
22            Feature:=.Cells(9, 1).Value, _
23            LapseRate:=.Cells(10, 1).Value, _
```

```
24                    ValDate:=.Cells(11, 1).Value, _
25                    NumSce:=.Cells(12, 1).Value
26       End With
27
28       MGMDB.InitializeScenario Sce:=Sheets("Base"), FR
            :=Range("FRBase")
29       dFMVBase = MGMDB.FMV(0)
30       dFMVEU = MGMDB.FMV(0.01)
31       dFMVED = MGMDB.FMV(-0.01)
32
33       MGMDB.InitializeScenario Sce:=Sheets("AllUp"), FR
            :=Range("FRAllUp")
34       dFMVIU = MGMDB.FMV(0)
35
36       MGMDB.InitializeScenario Sce:=Sheets("AllDn"), FR
            :=Range("FRAllDn")
37       DFMVID = MGMDB.FMV(0)
38
39       Sheet1.Range("O3").Value = dFMVBase
40       Sheet1.Range("O4").Value = (dFMVEU - dFMVED) /
            0.02
41       Sheet1.Range("O5").Value = (dFMVIU - DFMVID) / 20
42       Sheet1.Range("O6").Value = Timer - dStartTime
43   End Sub
```

A.11   Connecting to Databases

Solution 11.2. We can write the sub procedure as follows:

```
1    Sub TestCreateTable()
2        Dim sFileName As String
3
4        sFileName = ThisWorkbook.Path & "\va.accdb"
5        Connect (sFileName)
6
7        CreateTable Data:=Sheets("Main").Range("O2:Q5"),
            _
8            TableName:="Grade", PrimaryKey:="Name"
9
10       Disconnect
11   End Sub
```

Solution 11.4. We can write the sub procedure as follows:

```
1  Sub TestExtractRecord()
2      Dim sFileName As String
3      Dim sSQL As String
4      Dim vRes As Variant
5
6      sFileName = ThisWorkbook.Path & "\va.accdb"
7      Connect (sFileName)
8
9      sSQL = "SELECT * FROM Grade WHERE [Grade]='B'"
10     vRes = ExtractRecord(sSQL)
11
12     Disconnect
13
14     Dim i As Integer
15     Dim j As Integer
16     For i = 0 To UBound(vRes, 1)
17         For j = 0 To UBound(vRes, 2)
18             Debug.Print vRes(i, j) & " - ";
19         Next j
20         Debug.Print ""
21     Next i
22 End Sub
```

Executing the sub procedure gives the following output:

```
1  First Name - Carroll  -
2  Last Name - Paul -
3  Grade - B -
```

References

Bauer, D., Kling, A., and Russ, J. (2008). A universal pricing framework for guaranteed minimum benefits in variable annuities. *ASTIN Bulletin*, 38(2):621–651.

Black, F. and Scholes, M. (1973). The pricing of options and corporate liabilities. *Journal of Political Economy*, 81(3):637–54.

Bovey, R., Wallentin, D., Bullen, S., and Green, J. (2009). *Professional Excel Development: The Definitive Guide to Developing Applications Using Microsoft Excel, VBA, and .NET*. Addison-Wesley Professional, 2nd edition.

Boyle, P. and Hardy, M. (1997). Reserving for maturity guarantees: Two approaches. *Insurance: Mathematics and Economics*, 21(2):113–127.

Brown, R. A., Campbell, T. A., and Gorski, L. M. (2002). Valuation and capital requirements for guaranteed benefits in variable annuities. *Record*, 28(3).

Carmona, R. and Durrleman, V. (2005). Generalizing the Black-Scholes formula to multivariate contingent claims. *Journal of Computational Finance*, 9(2):43–67.

Cathcart, M. J., Lok, H. Y., McNeil, A. J., and Morrison, S. (2015). Calculating variable annuity liability "greeks" using Monte Carlo simulation. *ASTIN Bulletin*, 45(2):239–266.

Devroye, L. (1986). *Non-Uniform Random Variate Generation*. Springer, New York, NY.

Evans, M., Hastings, N., and Peacock, B. (2000). *Statistical Distributions*. John Wiley & Sons, Inc., Hoboken, New Jersey, 3rd edition.

Forta, B. (2012). *SQL in 10 Minutes, Sams Teach Yourself*. Sams Publishing, 4th edition.

Fu, M. C. (2015). Stochastic gradient estimation. In *Handbook of Simulation Optimization*, volume 216 of *International Series in Operations Research & Management Science*, pages 105–147. Springer, New York, NY.

Gan, G. (2011). *Data Clustering in C++: An Object-Oriented Approach.* Data Mining and Knowledge Discovery Series. Chapman & Hall/CRC Press, Boca Raton, FL, USA.

Gan, G. (2013). Application of data clustering and machine learning in variable annuity valuation. *Insurance: Mathematics and Economics,* 53(3):795–801.

Gan, G. (2015a). Application of metamodeling to the valuation of large variable annuity portfolios. In *Proceedings of the Winter Simulation Conference,* pages 1103–1114.

Gan, G. (2015b). A multi-asset Monte Carlo simulation model for the valuation of variable annuities. In *Proceedings of the Winter Simulation Conference,* pages 3162–3163.

Gan, G. and Lin, X. S. (2015). Valuation of large variable annuity portfolios under nested simulation: A functional data approach. *Insurance: Mathematics and Economics,* 62:138–150.

Gan, G. and Lin, X. S. (2016). Efficient Greek calculation of variable annuity portfolios for dynamic hedging: A two-level metamodeling approach. Revised and resubmitted to *North American Actuarial Journal.*

Gan, G., Ma, C., and Xie, H. (2014). *Measure, Probability, and Mathematical Finance: A Problem-Oriented Approach.* John Wiley & Sons, Inc., Hoboken, NJ.

Garrett, S. (2013). *Introduction to the Mathematics of Finance. A Deterministic Approach.* Butterworth-Heinemann, Waltham, MA, 2nd edition.

Garrett, S. (2015). *Introduction to Actuarial and Financial Mathematical Methods.* Academic Press, Singapore.

Getz, K. and Gilbert, M. (2001). *VBA Developer's Handbook.* Sybex, 2nd edition.

Gupta, A. K. and Varga, T. (2002). *An Introduction to Actuarial Mathematics.* Springer, New York, NY.

Hagan, P. S. and West, G. (2006). Interpolation methods for curve construction. *Applied Mathematical Finance,* 13(2):89–129.

Hagan, P. S. and West, G. (2008). Methods for constructing a yield curve. *WILMOTT magazine,* May/June:70–81.

ISDA (2006). 2006 ISDA definitions. International Swaps and Derivatives Association.

Lai, D. C. F., Tung, H. K. K., and Wong, M. C. S. (2010). *Professional Financial Computing Using Excel and VBA*. John Wiley & Sons, Inc., Hoboken, NJ.

Martin, R. (2012). *Excel Programming with VBA Starter*. Packt Publishing, Birmingham, UK.

Press, W., Teukolsky, S., Vetterling, W., and Flannery, B. (2002). *Numerical Recipes in C++: The Art of Scientific Computing*. Cambridge University Press, 2nd edition.

Richards, E. G. (2012). *The Explanatory Supplement to The Astronomical Almanac*, chapter 15, pages 585–624. University Science Books.

Roman, S. (2002). *Writing Excel Macros with VBA*. O'Reilly Media, 2nd edition.

Sheldon, T. J. and Smith, A. D. (2004). Market consistent valuation of life assurance business. *British Actuarial Journal*, 10(3):543–626.

Sintes, A. (2001). *Sams Teach Yourself Object Oriented Programming in 21 Days*. Sams, Indianapolis, Indiana, 2nd edition.

Tale, S. (2016). *SQL: The Ultimate Beginners Guide: Learn SQL Today*. CreateSpace Independent Publishing Platform.

The Geneva Association Report (2013). Variable annuities — an analysis of financial stability. Available online at: `https://www.genevaassociation.org/media/618236/ga2013-variable_annuities.pdf`.

Urtis, T. (2015). *Excel VBA 24-Hour Trainer*. Wrox.

Walkenbach, J. (2013a). *Excel 2013 Power Programming with VBA*. John Wiley & Sons, Inc., Hoboken, NJ.

Walkenbach, J. (2013b). *Excel VBA Programming For Dummies*. John Wiley & Sons, Inc., Hoboken, NJ, 3rd edition.

Index

Index of VBA Keywords

Printed in the United States
by Baker & Taylor Publisher Services